Daniel D. Fernández
The Man Behind the Medal of Honor

by

Richard Melzer, Cynthia J. Shetter, and John Taylor

A Valencia County Historical Society &
Los Lunas Museum of Heritage & Arts Publication

© Richard Melzer, Cynthia J. Shetter, and John Taylor
All rights reserved.
Valencia County Historical Society
Los Lunas Museum of Heritage & Arts
Los Lunas, New Mexico

Printed in the U.S.A.
Book design by Cynthia J. Shetter

No part of this book may be reproduced or transmitted in any form, or by any means, electronic or mechanical, including photocopying, recording, or by any information retrieval system, without permission of the publisher.

Names: Melzer, Richard, 1949 - | Shetter, Cynthia J, 1966 - | Taylor, John M. (John McLellan), 1947 - | Valencia County Historical Society (N.M.) | Los Lunas Museum of Heritage and Arts (N.M.)

Title: Daniel D. Fernández: The Man Behind the Medal of Honor / by Richard Melzer, Cynthia J. Shetter and John Taylor.

Description: Los Lunas, NM: Valencia County Historical Society & Los Lunas Museum of Heritage & Arts, 2023. | "A Valencia County Historical Society & Los Lunas Museum of Heritage & Arts Publication." | Includes bibliographical references and index.

Identifiers: LCCN 2023903192 | ISBN 979-8-9878437-0-3 (paperback: alkaline paper)

Subjects: LCSH: Fernández, Daniel D., 1944 – 1966. | Valencia County (N.M.) – History, Local. | Medal of Honor. | United States – Armed Forces – Biography. | United States – History, Military. | Vietnam Conflict, 1961-1975 – Regimental histories – United States.

Classification: LLC | DDC 355.1/342/0922

Cover: Daniel D. Fernández operating a 50-cal. M2 machine gun on an *M113* Armored Personnel Carrier.

Cover design by Cynthia J. Shetter

Dedication

To the Fernández Family

Acknowledgements

The authors gratefully acknowledge the support and assistance of Peter and Priscilla "Lollie" Fernández, Laurie Kastelic, Ray E. Sue, Hazel Willis, Louis Huning Jr., the staff at the Vietnam Veterans Memorial in Angel Fire, New Mexico, and numerous other individuals who willingly shared their remembrances of Daniel.

Contents

Introduction	1
Chapter 1: The War in Vietnam	3
Chapter 2: Growing Up	7
Chapter 3: Training and First Deployment	15
Chapter 4: Second Deployment and Final Patrol	53
Chapter 5: Coming Home	71
Chapter 6: The Medal of Honor	85
Chapter 7: Other Honors and Tributes	101
Chapter 8: Legacy of a Hero	145
Chapter 9: What Makes a Hero?	179
Epilogue: The Men Daniel D. Fernández Saved	183
Appendix 1: A Brief History of the Medal of Honor	193
Appendix 2: New Mexico Medal of Honor Recipients	201
Appendix 3: New Mexico's Medal of Honor Recipients Buried with Daniel Fernández at the Santa Fe National Cemetery	203
Appendix 4: Timeline	205
Appendix 5: List of Valencia County and Isleta Pueblo Residents Killed During the Vietnam Conflict	209
Bibliography	211
Index	219

Introduction

Los Lunas is one of the fastest growing communities in New Mexico. In the last 50 years, it has grown from a small village of 973 residents to a large town of 17,861. In exploring their new community, newcomers to Los Lunas discover many interesting places, including many named after famous former residents.

They might, for example, notice that a small monument in front of the middle school is dedicated to TSGT Foch Romero, killed in action during World War II. Or they might be curious about the name Solomon Luna, engraved above the entry to the town's original high school. Who was Fred Luna, for whom a senior center is named? In addition, who was Raymond Gabaldon, for whom a school is named?

Most of all, newcomers might wonder about the identity of Daniel Fernández, whose name appears at a large park, a Veterans of Foreign Wars post, and the community's newest bridge over the Rio Grande. As often happens, even the identity of someone as noteworthy as Daniel Fernández can fade over time.

Who was Daniel Fernández? When did he live? Moreover, what did he do to warrant such widespread, enduring respect and attention in his hometown and other parts of our nation?

It is the purpose of this book to answer these questions and to help all New Mexicans remember Daniel Fernández, one of the greatest heroes in state history. An early chapter of this book describes Daniel's youth, during which he developed his strong values and exceptional character.

Other chapters tell of his enlistment in the army, his training as an infantryman, and deployment to Vietnam in 1965. They also describe his experiences in Vietnam, on and off the battlefield, including a combat injury that led to his recuperation in Hawaii and back home in

Los Lunas. The book's most compelling chapter describes how Daniel heroically saved the lives of four of his fellow soldiers by sacrificing his own during his second deployment in Vietnam. Final chapters tell of his funeral and the many ways in which Daniel has been remembered in Los Lunas and far beyond. The concluding chapters reflect on his legacy and consider the qualities of a hero that Daniel exemplified leading to his ultimate act of self-sacrifice and bravery.

Daniel Fernández is deservedly remembered as a great hero and the only resident of Valencia County ever to receive the Medal of Honor. However, it is important to remember that Daniel was one of thousands of men and women from Valencia County who have served the United States and sometimes lost their lives in a long tradition of service, starting in the Indian Wars and continuing to recent conflicts in the Middle East.

In many of these conflicts, New Mexico has had a higher per capita rate of service than any other state in the Union. Always quiet and humble, Daniel would probably object to our telling his story without acknowledging all those who came before him and all those who continue to serve, including his nephew, Lt. Col. John Fernández, who proudly displays Daniel's Medal of Honor in his office at Schriever Space Force Base near Colorado Springs, Colorado. Daniel would have wholeheartedly approved of our dedication of this book in the memory of all those who have, like Daniel, given their lives to help protect their country, their values, and their loved ones back home.

Chapter 1: The War in Vietnam

Although Daniel Fernández enlisted in the U.S. Army in 1962, the roots of the conflict that brought him to Vietnam dated back to the 1870s when France exerted rather draconian control over what was then known as Indochina. The native Vietnamese, Cambodians, and Laotians resented being treated as second-class citizens and took every opportunity to throw off the burden of French colonialism.

During World War II, the Japanese asserted control over Indochina, but after their defeat in 1945, the French returned. A nationalist movement led by Ho Chi Minh, an avowed communist, opposed French colonial ambitions. Minh's ambitions were supported by both China and Russia. Western foreign policy, however, was dominated by the so-called domino theory, which asserted that if one nation fell to communism it was inevitable that one after another would fall until only the U.S. remained free. This meant that the only way to control the spread of communism was by containing it in place. Thus, the U.S. supported the French with military supplies as they opposed Ho Chi Minh and the nationalists.

Ho Chi Minh
(Texas Tech University)

After the French were decisively defeated at the Battle of Dien Bien Phu in 1954, a conference was convened in Geneva, Switzerland, to determine the future of Vietnam. It was agreed that Vietnam would "temporarily" be partitioned at the 17th parallel—the communists, under the leadership of Ho Chi Minh, would control the north, and a democratic government, under the leadership of Ngo Dinh Diem, would control the south.

Despite the violence and corruption of the Diem regime, the United States felt obligated to support the regime as a buttress against the communist north. In 1959, the government of North Vietnam secretly decided to support a liberation movement in the south. This movement, known as the National Front for the Liberation of South Vietnam, or Viet Cong, began a campaign of guerilla warfare against Diem. As a reaction, in 1959 the first contingent of 760 American troops were deployed to Vietnam as military advisors.

Ngo Dinh Diem
(Department of Defense)

As the fighting in Vietnam increased, US support in terms of military equipment and advisors tried to keep pace. President John F. Kennedy and later President Lyndon B. Johnson reiterated the U.S. commitment to fighting communism in Vietnam. By the time Daniel Fernández enlisted in the Army in 1962, the size of the U.S. military contingent in the region had grown to 11,300.

The flag of the Viet Cong, adopted in 1960, is a variation of the North Vietnamese flag. (Wikimedia Commons)

The flag of the Republic of Vietnam was designed by South Vietnamese painter Celso-Leon Le Van De in 1948. (Wikimedia Commons)

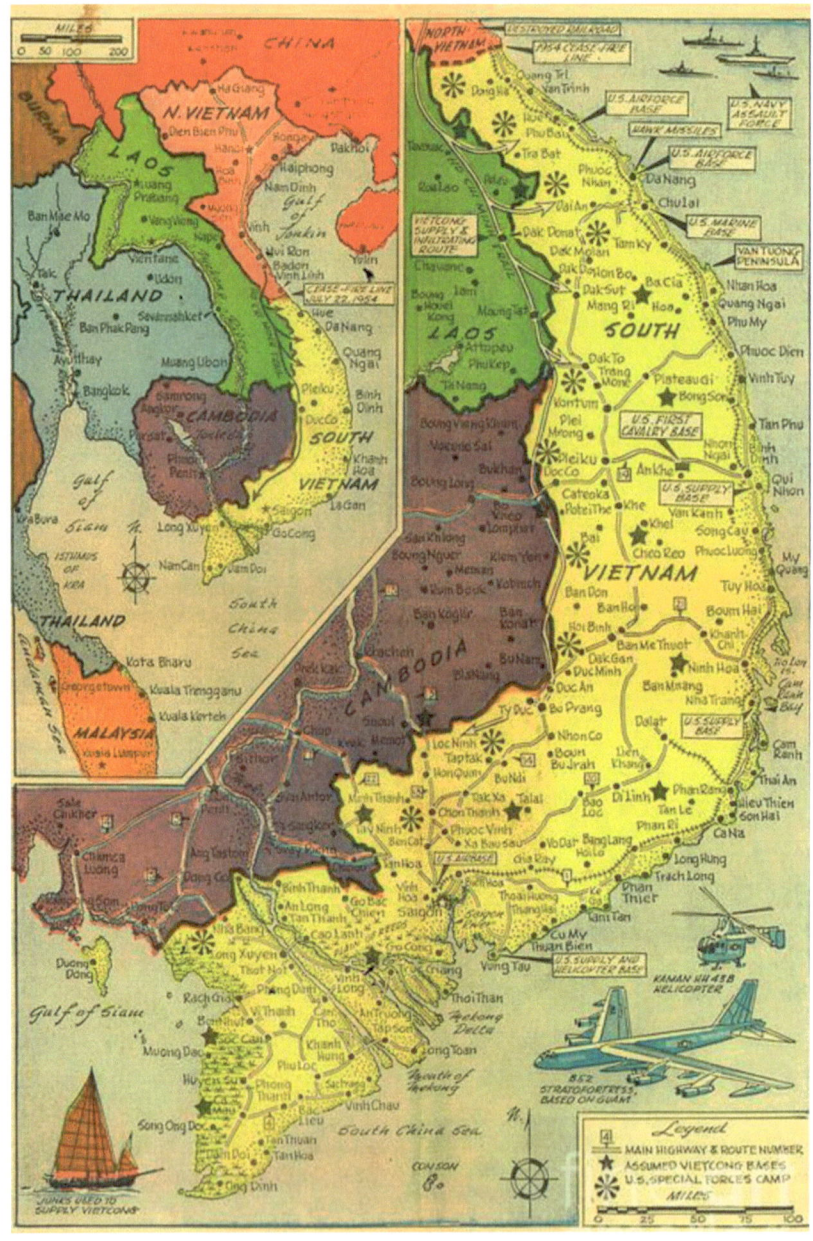

Map of Vietnam (Dallas Morning News)

Chapter 2: Growing Up

Daniel D. Fernández was born at Presbyterian Hospital in Albuquerque, New Mexico, on June 30, 1944, to José Ignacio Fernández and Lorinda "Laura" Griego Fernández, joining his two older brothers Joseph Anthony, born July 20, 1942, and Frank David, born one year later on July 12, 1943. He was baptized Daniel Damaso Fernández at Sacred Heart Catholic Church in Albuquerque, New Mexico, on September 10, 1944. Soon after Daniel's birth, his mother Laura contracted tuberculosis, as did his two brothers.

Laura's mother, Angelita Telles Griego, grew up near Fort Stanton, New Mexico, where the government had converted the fort into a tuberculosis sanatorium. Angelita knew that rest and good nutrition were the prescribed treatments for the devastating lung infection. She watched Daniel while Laura and the boys recuperated. Tragically, the disease progressed, causing Laura to have a lung removed and Daniel's brothers Joseph and Frank to succumb to the infection in September and October of 1944. These staggering deaths were immediately followed by the loss of another infant son, Victor, in 1945 at 12 days old.

Healing from the overwhelming loss of three children in the span of a year, Laura was comforted by her

José Ignacio Fernández
(Fernández Collection)

husband José and her one-year-old son, Daniel. Shortly, Laura became pregnant once again, and began praying to Santa Rita, the patron saint of the Catholic church she attended while growing up in the farm and ranch community of Carrizozo, New Mexico. On July 10, 1946, Laura and José welcomed a newborn girl they named Rita after the patroness of heartbroken women. Laura's health continued to improve, and on June 20, 1949, they welcomed Peter to the Fernández family.

The Fernández family made their home in Albuquerque until José had the opportunity to work for the United States Department of Agriculture in Guadalajara, Mexico, in 1949. José had been born on a ranch in northern New Mexico between Chico in Colfax County and Gladstone in Union County. His family raised sheep and cattle and engaged in farming. This knowledge of agriculture, along with his organizational skills and fluency in Spanish, benefited José as he aided Mexican ranchers in the eradication of hoof and mouth disease that was ravaging livestock in Mexico.

Laura and the children joined José in Mexico. Daniel, Rita, and Peter attended school where they spoke fluent Spanish. In 1952, the Fernández family moved back to Albuquerque. Daniel attended Cortez and Queen of Heaven schools. He and his siblings only spoke Spanish, having lived in Mexico for three years. Rita and Peter were able to make the transition more easily

Lorinda "Laura" Griego Fernández (Fernández Collection)

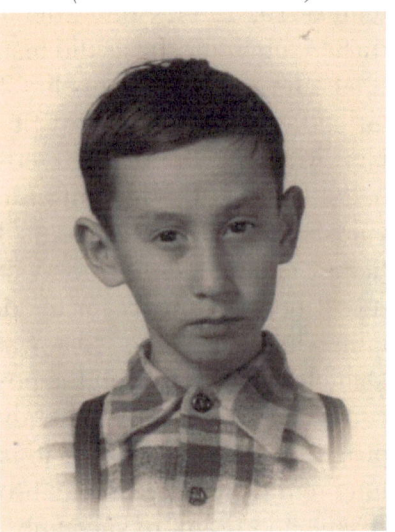

Daniel D. Fernández (age 8 or 9) (Fernández Collection)

than Daniel. The language barrier caused Daniel to be held back a year, putting him and Rita in the same grade.

Young Daniel joined the Cub Scouts, and his mother was the den mother. His father, José, worked for a title abstract company where he traveled to Sandoval, Bernalillo, and Valencia Counties. The real estate title business started growing rapidly due to growth in the eastern part of Valencia County and uranium mining in the western part of the county. José and Laura decided to move to Los Lunas where José could work in his company's branch office. The family expanded once again with the birth of James Henry on February 26, 1955. In August 1955, Laura and José purchased a three-acre farm located in Chihuahita, a neighborhood south of the Village of Los Lunas. The farm was close to the county courthouse where they both worked. Here they planned to raise their own food and give their children the rural farming atmosphere they had experienced as children.

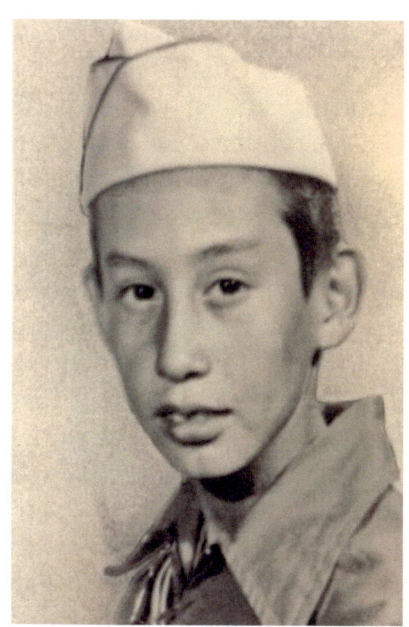

Daniel in his School Safety Patrol Uniform (Fernández Collection)

Daniel was 11 years old when the family moved to Los Lunas. For Daniel their new farm became a haven. They raised sheep, calves, chickens, and pigs. They also grew alfalfa and had an apple orchard. Daniel grew to love this rural life and joined 4-H to learn more about farming and ranching.

Daniel joined the School Safety Patrol, a program sponsored by the American Automobile Association (AAA). Wearing his white diagonal Sam Browne belt, Daniel would ensure that boys and girls walking to school stayed on the safe side of the curb at regular crossings, until the traffic allowed them to cross.

Daniel often visited his Uncle Louis Fernández on his ranch near Springer, New Mexico. "We had to cross the Cimarron River on horseback," said his father, José. "Sure, that was dangerous. There were big holes in the river, but Danny didn't hesitate." Danny didn't hesitate about many things.

Charles Sullivan tells of one such story. Charles and his brother Kenneth, known as "Tay" moved into the Luna Mansion about the same time that Daniel moved to Los Lunas. Charles said that one day a group of boys jumped him and Tay. Daniel ran into the midst of it and saved them from getting beat up too badly, making them fast friends.

One day, right after Christmas, Charles was in his front yard shooting a bow and arrow when Daniel rode up on his bicycle. Daniel had his brand-new BB gun with him. Charlies and Daniel got to shooting, and as Ralphie's mother from the movie "A Christmas Story" predicted, Daniel accidently shot Charles in the eye. There was no permanent damage, but Daniel sure got an earful from his parents.

Daniel's Uncle Louis Fernández (Ancestry.com)

Daniel, Charles, and Tay went horseback riding all over Los Lunas. Charles chuckles at one memory of their time as kids. One day they were pretending to be capturing prisoners. Tay was on horseback leading Daniel, the prisoner, on foot with his wrists tied together with a rope. Tay's horse got to acting up and took off running. They dragged Daniel several yards through the weeds by his wrists before Tay was able to let loose of the rope.

Daniel's friend, Cruz Muñoz, remembers climbing up El Cerro de Los Lunas with Daniel and three other buddies. Always daring, the boys ran down, jumping over plants in their way. Daniel cleared one such bush only to find a large cliff on the other side. He barely hung on to a ledge as his friends managed to pull him out of danger.

Daniel became known for his ability to break horses in a gentle manner. In an interview, Rebecca Richins-Varela said, "Daniel was a horse whisperer. He would take his time to tame a horse, often sleeping with it upon the ground." He was a calm, gentle person who was more into taming than competing. He spent a lot of his time in Bosque Farms where he worked with a two-year-old filly. He was known for his ability to ride horses that nobody else would touch.

County Tax Appraiser M.S. Romero recalled after Daniel's death: He was a terrific bronc buster, that boy. He broke that filly [named Lady] and he did it without breaking her spirit... It was right here in this field.... He had a kid's saddle on her and didn't even have his feet in the stirrups. The saddle slipped up around her neck and he just slid backwards and broke her bareback.... You know the first thing he did when he came home.... He bought a pair of boots and went riding.

"He was always dressed like a Texan – you know, jeans and boots all the time," said Willie Gómez, custodian at the courthouse. "I give him credit. He would ride horses nobody else would touch. He had the guts to do it."

Daniel D. Fernández's hand tooled leather belt. (Fernández Collection)

Daniel could harvest honey from the neighbor's hives without smoke or wearing protection. He was such a calm, gentle person that the bees wouldn't sting him. He sold apples from the family orchard. "I remember that Daniel was a good salesman. He and his brother, Peter, would go out in the afternoons to sell the apples and the fruit buckets always came back empty," said his mother. "With the money he made from selling apples and the animals, he would buy the western style clothing he loved." These activities helped foster his dream of owning a horse ranch someday.

Daniel was active in sports as well, especially track and cross-country. He was a tall, wiry, lanky runner; his friends said he galloped like a horse, earning him the nickname *El Caballo*. Daniel was fast. He would also hike, ride bicycles, and play pick-up football with his buddies the "49ers."[1] Frank Gurulé recalls that he tried to recruit Daniel for the high school football team because he could "take a hit," but Daniel just wasn't interested in "taking orders" from a coach.

---

[1] No one seems to know where the moniker "49ers" originated. Frank Gurulé, one of the group, says it definitely wasn't a reference to the San Francisco football team. Frank recalls that they were "sort of like the Little Rascals" of television fame. There were 12 members in the group, and they hung together from elementary through middle school. The "Dirty Dozen" included Larry, Louis, and Salo Artiaga, Gilbert and Walter Baca, Nick Balido, Art and Randy Castillo, Tony Gómez, Frank Gurulé, Cruz Muñoz, and, of course, Daniel Fernández.

Always willing and anxious to help, Daniel prevented a fire from starting in the house of his neighbor, Rosalie Márquez. When an electrical short threatened to start a fire, "Danny rushed in and turned off the electricity, then started carrying things out."

*Daniel and Rita Fernández, junior year. (*La Luna *1963)*

It was while running cross-country that Daniel met Rebecca "Becky" Richins (now Varela). She was three years younger than he was. Rebecca often accompanied "Dan D," as she and friends in their group called him, to the Fernández farm where he worked with his horse and other animals. "He was a kind, gentle man who helped the elderly ladies in the neighborhood with their animals, often taking the animals to the veterinarian for them," Becky reflected.

Daniel D. Fernández on left. (La Luna *1962*)

Becky moved to Lordsburg in 1962, during her freshman year, but they kept in contact.

Daniel was an average student, but school just didn't interest him. In his junior year in high school, he decided to drop out of school to join the service. Becky tried to change Daniel's mind, but he felt it was his civic duty. Rebecca Lutz, his 11th-grade English teacher, also tried to convince him to finish high school before he enlisted. However, he was determined, saying that if he was going to enlist, he might as well

*Rebecca Richins (*La Luna *1962)*

do it then. He later joked that he flipped a coin with the recruiter and lost, but, in fact, he voluntarily enlisted. He received his basic training at Fort Polk, Louisiana.

*Daniel Fernández's official military portrait, 1963
(Fernández Collection)*

Chapter 3: Training and First Deployment

After completing basic training at Fort Polk, Daniel was assigned to Schofield Barracks in Hawaii. He lived up to a promise to his English teacher by completing his GED while he was going through intensive jungle training.

Daniel's GED Certificate (Fernández Collection)

Daniel participated in training exercises in the Hawaiian Islands that concentrated on jungle fighting reminiscent of the bitter battles on Guadalcanal in 1942 and the historic Philippines Campaign of World War II.

The 25th Infantry Division believed that a soldier cannot fight in the jungle until he learned to survive in the jungle. Soldiers must learn to use the jungle as a friend – not as an enemy – to become a skilled jungle fighter. The 25th Infantry proved this by training every man at a unique Special Asian Warfare Training and Orientation Center (SAWTOC) that was located 25 miles from Waikiki's beaches in Hawaii. The soldiers received training from instructors that they had received during training in Southeast Asia during Southeast Asia Treaty Organization (SEATO) exercises in Thailand.

SAWTOC was located in the rugged, jungle-like terrain in the foothills of Oahu's Koolau Mountains. The center prepped soldiers at nine stations designed to teach them first to survive in the jungle and, second, to fight in the jungle. These fighting techniques were taught at the Jungle and Guerrilla Warfare Training Center or Jigwits as the soldiers called it.

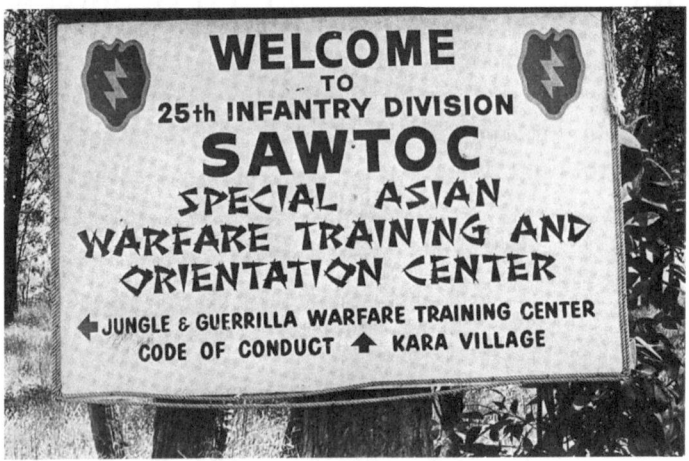

Special Asian Warfare Training and Orientation Center (SAWTOC) on Oahu. (25th Infantry Division)

Every man in the 25th Infantry Division was required to train at the facility at least once a year. The course of instruction lasted five and one-half days and was divided into two phases, consisting of instruction and practical application under combat conditions in the field.

During the first phase, Daniel went through each of the nine stations. He learned mountain climbing and rappelling from cliffs and helicopters. He learned how to cross bodies of water with his weapons and equipment on ropes and to build rafts from his own poncho, trousers, coconuts, ammunition boxes, or any other material. Daniel learned to navigate by day and night through seemingly impenetrable jungles. At other stations, he learned guerilla tactics like how to set an ambush and to counter ambushes. He learned about aerial resupply and evacuation.

Daniel was taught first aid and the importance of hygiene in the jungle. Keeping his feet dry was of the utmost importance. He developed field craft abilities such as how to build a shelter with available materials. He acquired survival skills in how to obtain food and water in the jungle, to trap fish, birds, and animals as well as how to distinguish between edible and poisonous plants. Daniel also gained knowledge in how to obtain water and oil from coconuts and to weave with grasses and leaves.

The second phase of the training was called Operation Bushwacker, a two and one-half-day field exercise during which the entire company was required to fight its way through guerilla-occupied territory, utilizing information given during the first three days of training.

SAWTOC Rappel station. (25th Infantry Division)

Another element of Daniel's training involved learning how to control the grip the adversary had on local civilian populations. The soldiers of the 25th prepared for this guerilla type of warfare in a bustling replica of "anywhere" Southeast Asia, complete with inhabitants in Kara Village, located at Schofield Barrack's East Range.

The food, clothing, and sometimes dialect were as real as it could be made. The complex was patterned after a village in the upland portion of the Republic of Vietnam. Kara Village was used as a training vessel to instruct the men of the 25th Infantry Division in the methods and techniques that stressed the Army's civil affairs program of assisting the local population in defending themselves, improving their standards of living, and establishing a working relationship in the procedures employed when entering a strange village.

General William Westmoreland's wife, Katherine (Kitsy), spent a year in Vietnam before she moved to Hawaii when dependents were sent out of the war zone. She stated, "The training village looks like the real thing. Except, of course, you can never capture the smell, or the poverty, of Southeast Asia."

Aerial view of Kara Village in Oahu's Koolau Mountains. (25th Infantry Division)

These thatched-roof huts clustered in four villages in the Koolau Mountains were products of division ingenuity. The "native inhabitants," specially trained 25th reconnaissance troops, created realistic situations for the division's jungle training exercises. (25th Infantry Division)

 Jim Becker with the *Honolulu Star-Bulletin* observed an exercise in August of 1965. He stated there were soldiers acting as Vietnamese villagers wearing black peasant clothing that resembled pajamas and conical straw hats sitting around as an American unit surrounded their village. During the exercise, the American soldiers' mission was to search for enemy soldiers and weapons.

 The soldiers started their search at a chicken coop, where they were met by hysterical squawks from the maddest hens Becker ever saw. There was nothing there, but they found a hidden rifle two huts down. When they got to the schoolhouse, a Viet Cong soldier hiding there threw a grenade in the middle of the square and ran. The grenade went off, but the American soldier "shot" the Viet Cong soldier. The soldiers moved to another hut, where they found a tunnel. In went a smoke grenade. Ten seconds went by, and a mock Viet Cong soldier popped out of the well, choking from the smoke.

There was also Code of Conduct training where Daniel and other soldiers were taught carefully worded scripts based on Executive Order 10631, signed by Dwight D. Eisenhower on August 17, 1955, in the event they were captured.

The Code of Conduct for Members of the Armed Forces of the United States:

I. I am an American fighting man. I serve in the forces, which guard my country and our way of life. I am prepared to give my life in their defense.

II. I will never surrender of my own free will. If in command, I will never surrender my men while they still have the means to resist.

III. If I am captured, I will continue to resist by all means available. I will make every effort to escape and aid others to escape. I will accept neither parole nor special favors from the enemy.

IV. If I become a prisoner of war, I will keep faith with my fellow prisoners. I will give no information or take part in any action, which might be harmful to my comrades. If I am senior, I will take command. If not, I will obey the lawful orders of those appointed over me and will back them up in every way.

V. When questioned, should I become a prisoner of war, I am bound to give only name, rank, service number, and date of birth. I will evade answering further questions to the utmost of my ability. I will make no oral or written statements disloyal to my country and its allies or harmful to their cause.

I will never forget that I am an American fighting man, responsible for my actions, and dedicated to the principles which made my country free. I will trust in my God and in the United States of America.

These Code of Conduct trainings were taught in conjunction with Prisoner of War (POW) organization and the principles of how to plan and carry out an escape and evasion. The four-hour block of instruction demonstrated conditions known to exist in various POW camps.

Sign at the Mock POW training camp at SAWTOC. (25th Infantry Division)

Mock POW Camp used at SAWTOC to train soldiers in the event of capture. (25th infantry Division)

Operation Shot Gun

While stationed in Hawaii, Daniel and a few of his friends had an opportunity to volunteer for additional training for an elite group deployed to Vietnam as part of a classified operation.

On August 6, 1964, the 3rd Aviation Company Airborne Multi-Intelligence Laboratory (AML) was activated and assigned to the 3rd United States Army for eventual movement to Vietnam. The Company was attached, for logistical and administrative support, to the 11th Air Assault Division at Fort Benning, Georgia. The 3rd Aviation Company received an equipment readiness date of October 21, a personnel readiness date of November 27, and the Army Post Office (APO) address of Bien Hoa, Vietnam, on December 4, 1964.

Upon arrival in Vietnam, approximately 75 percent of the 3rd Aviation Company was integrated with personnel from other companies of the 145th Aviation Battalion. The company was housed in Bailey Compound at Bien Hoa with the 118th Assault Helicopter Company, known as the Thunderbirds. On December 14, 1964, the 3rd Aviation Company (AML) was re-designated as Company A/501st Aviation Battalion. On December 26, 1964, Lt. Col. Robert K. Cunningham, CO, 145th Aviation Battalion, declared the Company operational.

The Company and platoons selected new call signs for use in Vietnam. Major Lewis J. Henderson's wife, Ramona, gets credit for the Firebird call sign (from Stravinsky's ballet). Captain Cliff Ohlenburger designed the emblem/patch for the Firebird gunship platoon.

U.S. Army Company A/501st Aviation Battalion Firebirds patch. (Fernández Collection)

Daniel and several members of the 25th Division stationed at Schofield Barracks joined the Firebirds in Vietnam on November 27, 1964, where they were to participate in the classified maneuvers.

Despite incredible danger, Daniel volunteered to be a "Shotgun Rider" on U.S. Army helicopters for 90 days in Vietnam. He was trained as a door gunner on a Bell UH-1 Iroquois (nicknamed Huey) and assigned to Company A of the 501st Aviation Battalion of the 3rd U.S. Army, the Firebirds, where he participated in more than 25 aerial missions over hostile territory before being injured during a mission.

In an interview with Hazel Willis, Hazel stated that "Danny" Fernández was a member of "Shot Gun IX" along with her brother Dan R. "Danny" Shearin. "Operation Shot Gun" was developed because the U.S. government did not want to acknowledge any kind of troop buildup in Southeast Asia. The operation was initially classified as Top Secret.

Chapter 3: Training and First Deployment

The highly classified "Shot Gun" program started in early 1962 after the Military Assistance and Advisor Group (MAAG) requested assistance with this issue. The Military Assistance Command, Vietnam (MACV) requested combat-trained men to take over the job of door gunners to man automatic weapons to protect the helicopters on operational missions. The plan was for the 25th Infantry Division in Hawaii to train and send gunners from Hawaii for short periods of time, roughly 90 days. This program was called Provisional Machine Gun Platoons that became known as "Shot Gun Platoons" or "Shot Gun Riders" much as you would see guarding stagecoaches in the Old West.

From its inception, the Shot Gun program was entirely composed of volunteers. Young officers with the proper security clearance were advised of the program and offered an opportunity to volunteer for 90-day, overseas Temporary Duty (TDY) assignments, in command of soldiers that they would interview, select, and train.

SFC (Ret.) Frankie Willard, who served in Company C, 2nd Battalion, 35th Infantry from 1961 to 1967, recalled how volunteer enlisted gunners were approached. He stated that men interested in volunteering were invited to a morning meeting at the Schofield Barracks main post theatre. Once all were inside and seated, the 25th Infantry Division Commander took the stage and said words that were "forever embedded" in Willard's mind: "Thanks for coming, but before I get started, I have something to say. Some of you will be selected to go in harm's way, and you could be killed. If you do not desire to go, I will wait for you to get up and leave right now." The General stopped talking and Willard notes that quite a few did get up and leave. The theatre doors were then closed, and Military Policemen were posted before the Division G-2 and G-3 gave the men a classified briefing on the tactical situation in Vietnam and the training that they would receive, if selected.

As Shot Gun Riders, Fernández and Shearin had to pass a Class III flight physical examination in which vision, color blindness, hearing, and other physical conditions were checked. Training was primarily with the M-60 machine gun, but volunteers also were expert with the .50 caliber machine gun, the M79 grenade launcher, the .45 caliber pistol, the M3 machine gun, the .30 caliber M2 carbine, and the M16 rifle and the M14, complete with bayonet.

While training in Hawaii, the platoons were first introduced to helicopter flight in a mock-up. They were oriented to the various gun mounts and in free firing with a "bungee cord." A bungee cord or strap would be mounted in the top of the gun door and the machine gun hooked to it for support.

Training also included techniques of aerial observation and firing from various altitudes; in addition, how to respond with instantaneous but planned reactions. Accuracy was constantly emphasized, especially in the descent to a landing zone.

Fernández and Shearin were in the 9th Provisional Machine Gun Platoon (the 9th group of men trained and sent to Vietnam) with the operational name Shot Gun IX.

A Shot Gun Rider with M-60 machine gun attached to a bungee cord. (Daniel Fernández Photo Albums)

In May of 1965, after Fernández and Shearin had returned from Vietnam, the two-year media ban was lifted on the highly classified operation. Bob Jones, a military reporter for *The Honolulu Advertiser*, broke the story of Operation Shot Gun and was invited to ride along with Shot Gun X as they trained.

Jones reported that from February 1963 to May 1965, nine soldiers from Schofield Barracks had died in South Vietnam and 96 were wounded in Operation Shot Gun. The 25th Infantry Shotgunners had seen more combat action than any other U.S. forces in Vietnam and were the only American troops who had a "license to kill," meaning they had permission to engage with communist troops.

The training Shot Gun X platoon received was similar to the training Fernández and Shearin received with Shot Gun IX. Jones described some of the training he witnessed:

> Yesterday, the GIs were getting live fire practice, learning how to engage an enemy with deadly fire from an M-60 machine gun carried on a special bracket just outside a helicopter's right door.
>
> It is the last phase of 112 hours of intense preparation before the men are assigned to scattered aviation units in Viet Nam.
>
> The morning program included low-level firing from a UH-1B

copter at McCarthy Flats, flying 50 feet off the ground and shooting at yellow barrels on a hillside some 2,100 feet away.

Firing the machine gun accurately as the copter moves along at a brisk pace requires a keen eye and sense of distance. Those who don't have these talents are washed out of the program.

Most of the trainees know the job is hazardous because in Viet Nam they are fully exposed to ground fire as they ride in the helicopter's door during landings and take-offs. But they don't seem to mind, and some are even going back for a second duty tour.

"It might be the glamour of war, but they all have to have motivation," said Major Gene P. Tanner of Bowling Green, Ky. [sic], the man in charge of training them.

The Schofield men had afternoon practice yesterday in Makua Valley, firing away from 1,000 feet at old trucks and barrels on the ground below them.

Already they had learned some unusual things.

Each man must be able to reassemble the scrambled parts of two types of rifles while blindfolded in five minutes.

Each man has been through a course in medical treatment and survival, a useful skill if his helicopter is shot down.

When their current 52 hours of actual aerial gunners is finished, the shotgunners will be off to Viet Nam. And another 400 volunteers are standing by.

In a speech before the U.S. Senate, Senator Daniel K. Inouye of Hawaii said:

Within the 25th Infantry Division there is an elite corps of officers and men carrying on the finest traditions of the American fighting man. They are called shotgunner by friends, and a far worse name by the Viet Cong they volunteer to fight. They know who the enemy of world freedom is – they believe in what they are doing – and as the 25th Infantry Division motto states, they are "Ready to Strike, Anywhere! Anytime!"

By late 1965, when the 25th Division was fully deployed to Vietnam, more than 2,000 officers and men from the "Tropic Lightning" Division had completed the training and gone for temporary duty to Vietnam.

Schofield Barracks in Hawaii, 1965. (Hazel Willis)

Daniel Fernández at Schofield Barracks in Hawaii. (Larry Artiaga)

Chapter 3: Training and First Deployment

Besides its reconnaissance and transport functions, the helicopter acquired a new role in Vietnam. The helicopter became a weapons platform, a gunship that could seek out and destroy enemy targets.

A bold and imaginative development, the concept of the helicopter assault appealed to American military tacticians because it enabled them to shift forces rapidly and on short notice. In this context, the aircraft assumed the role of the light mechanized columns in hit-and-run raids and armed reconnaissance operations.

The Army described the helicopter's new mission as providing unprecedented mobility during the Vietnam Conflict. Without the helicopter it would have taken three times as many troops to secure the 800-mile border with Cambodia and Laos.

The average infantryman in Vietnam saw about 240 days of combat in one year thanks to the mobility of the helicopter, compared to an infantryman in the South Pacific during World War II who saw about 40 days of combat in four years. Approximately 12,000 helicopters saw action in Vietnam. Army UH - 1s totaled 9,713,762 flight hours in Vietnam between October 1966 and the end of American involvement in early 1973. Army AH - 1Gs totaled 1,110,716 flight hours in Vietnam. Helicopters in Vietnam seldom flew above 1,500 feet, which is traffic pattern altitude for bombers, leaving them exposed to hostile fire even in their base camps. The average lifetime of a helicopter gunner in Vietnam was two weeks.

Shot Gun IX heading to Vietnam from Hawaii in November 1964 to join the 501st Aviation - Firebirds. (Hazel Willis)

Shot Gun IX arriving in Vietnam on November 27, 1964. (Hazel Willis)

Shot Gun IX retrieving equipment and belongings after arriving in Vietnam, November 27, 1964. (Hazel Willis)

Daniel operated a 60-M machine gun and a .50 caliber gun mounted as part of the armament system of the helicopter. (Daniel Fernández Photo Albums)

Daniel posing with a rocket that he rearmed helicopters with after missions. (Daniel Fernández Photo Albums)

Daniel D. Fernández: The Man Behind the Medal of Honor 29

UH-1 "Huey" helicopter on tarmac. (Daniel Fernández Photo Albums)

Helicopter gunships in dry rice fields. Soldiers sitting in shade. (Daniel Fernández Photo Albums)

A Shot Gun Rider in a UH-1 "Huey" helicopter. The M-60 machine guns are strapped or bungee tied to the top of the gun door frame. (Daniel Fernández Photo Albums)

SGT Dan R. Shearin, North Carolina, member of Shot Gun IX. (Daniel Fernández Photo Albums)

Emergency Medical training course certificate Daniel received while training for the Shot Gun Program at Schofield. (Daniel Fernández Photo Albums)

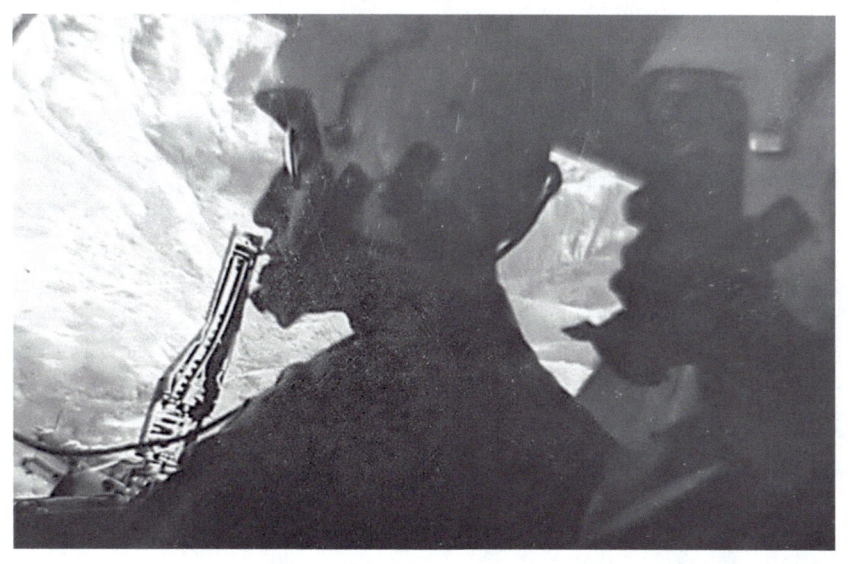

Daniel shown in flight with .50 caliber machine gun. (Daniel Fernández Photo Albums)

32   Chapter 3: Training and First Deployment

Daniel sitting in a helicopter on metal tarmac fabricated by engineers in Vietnam. (Fernández Collection)

Daniel standing in the helicopter gun door holding a 60-M machine gun. The gunship's rocket armament can be seen mounted at the lower left of the door. In flight, Daniel would be seated next to the rockets when they were fired. Often he would simultaneously be firing his weapon at enemy troops. The helmet was installed with a communication system to help him relay other aircraft, troops, and enemy positions to the pilot and crew.
(Laurie Kastelic Collection)

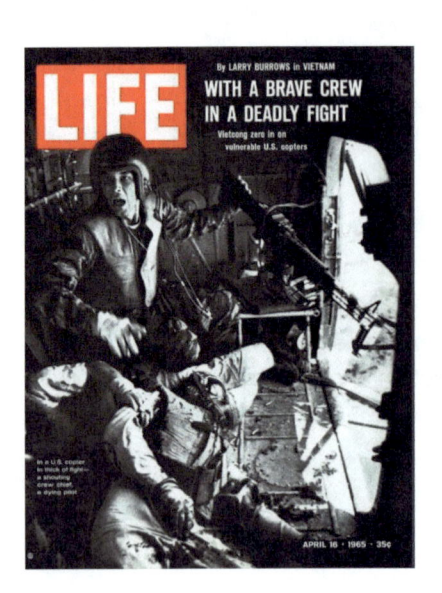

The peril Daniel faced is depicted in this April 16, 1965, photo on the cover of Life Magazine, showing a helicopter gunner and mortally wounded pilot. (Life Magazine)

These hovering U.S. Army helicopters depict the conditions Daniel faced when he was injured in his first tour of duty in Vietnam. Here gunships pour machine gun fire into a tree line to cover the advance of South Vietnamese ground troops in an attack on a Viet Cong camp 18 miles north of Tay Ninh, near the Cambodian border, in March of 1965.
(Pulitzer Prize Winner, Horst Faas/AP)

An Army helicopter flies in support of ground forces during the Battle of Ia Drang in November 1965. This is an example of the support Daniel and his flight crew provided for South Vietnamese ground troops during his first tour in Vietnam. (Department of Defense VIRIN:65114-O-ZZ999-001)

34 Chapter 3: Training and First Deployment

Daniel riding a motorcycle in his down time.

Daniel and a buddy lounging in the shade of truck listening to a radio.

Daniel (center) playing horseshoes. (Photos: Daniel Fernández Photo Albums)

The soldiers of the 25th Division were oriented on Vietnam and its customs and traditions before being deployed. Here is an article from the *Tropic Lightning News Orientation Edition:*

"Vietnam — Its Customs & Traditions"

The customs and traditions practiced by the Vietnamese people, in whose country you are a guest during your tour with the Free World Forces, may be traced back through many centuries.

The people are courteous, patient, sensitive, and quiet. To get along with them well and to mix with them easily and courteously it is important to gain an understanding of their way of life.

It is crucial that American servicemen cultivate an awareness of the Vietnamese culture and appreciation of the aspirations of the people. Begin now!

Here are some of the do's and don'ts when dealing with the people.

Vietnamese Home

The Vietnamese regards his home differently than does a westerner. He gives all the loyalty, concern, interest, and respect to his home that Americans feel for their community nation and entire earth.

His home is the focal point of his life. All other considerations are secondary to the family and home. Yet he considers himself a furnishing in his home, temporary and unimportant in the long run of time.

The Vietnamese code of behavior fosters courtesy, patience, and kindness. Do not mock it as it will negate the objectives of the assistance we are giving the people.

Taboos Familiar

Some of the taboos are familiar to readers of Emily Post or Dear Abby. They include:

Do not put your feet on any part of the furniture; do not be loud or overly emotional in public (Vietnamese of opposite sexes will seldom hold hands in public); do not ask about the price of personal possessions; do not shove your opinions or advice upon others unless it is asked for; respect those older than yourself.

The Vietnamese practice these rules of etiquette much more than Americans and regard violations seriously.

Examples Cited

Some taboos offend both the Vietnamese and his family, perhaps even his ancestors.

Never pat anyone on the head and never cross your legs so that your feet point toward anyone or to a shrine. These are based on the philosophy of humiliation and veneration.

It is a good rule to avoid gregarious "convention" manners in Vietnam. One should not offer to shake hands unless the Vietnamese does so first. Do not be a back slapper.

Vietnamese people do not permit use of their first names lightly. Call each by his rank or the proper form of the polite address. Incidentally, when writing a Vietnamese name, spell it out in full.

Beware of Gestures

Gestures are a booby trap for Americans in Vietnam. Each gesture has so many variations between our intent and a citizen's interpretation that trouble may arise unwittingly. Save hand and arm signals for Americans only.

At the dining table, do not eat until the eldest present has started. Clean your own plate but not the serving platter. If you repay hospitality, do not take him to a "hamburger joint" even though the food may be better. Status is important. The senior in age always picks the tab up.

If you send gifts to a Vietnamese, send them to the children if possible. Never send an odd number. Two boxes of candy are much better than one.

Differences Endless

The number of differences goes on and on, but the basic difference is one of degree. The Vietnamese are much more formal and traditional, almost exaggerated, in observing the amenities of life because they believe everything is integrated since the beginning of time.

To break or mock these culture codes is to invite disgrace and disaster by offending God, ancestors, and nature.

If the American serviceman thinks before he speaks or acts, he will save unnecessary grief. Common sense, courtesy, and application of the Golden Rule will help him steer clear of difficulties.

While "in country" Daniel developed a deep respect for the Vietnamese people as well as for the soldiers whom he fought alongside. Daniel's brother James summed up Daniel's affinity for all people:

> He easily made friends with people of all walks of life. One time the family went to the Laguna Fiestas. When it came time for some private ceremonies, they were asking the Anglo people to leave. My parents had to retrieve Daniel, dressed in his typical jeans, boots, and western shirt, from among the pueblo peoples because they thought he was one of their own.

On reflection, his brother James said that Daniel was comfortable amongst many different types of people. The more downtrodden, the more affinity he had with them. In one instance, a friend of Daniel's was having difficulty with his parents, so unbeknownst to his parents, Daniel decided to put his friend up in the old outhouse on their farm. That evening, his father heard a noise and went to investigate. He was quite surprised to find a boy set up for the night in his old privy.

Daniel's compassion for people and their way of life is reflected in the pictures he took during his first tour to Vietnam. Taking pictures of their daily life, he made friends with the South Vietnamese soldiers and learned some of the language through his training and assimilation during the short time he was there.

Daniel shaking hands with South Vietnamese soldiers. (Daniel Fernández Photo Albums)

A typical Vietnamese farmer's hut with a young child in the doorway. (Daniel Fernández Photo Albums)

Homes were bamboo structures with clay walls and thatched roofs. (Daniel Fernández Photo Albums)

A Vietnamese street market. (Daniel Fernández Photo Albums)

A South Vietnamese soldier and a civilian in front of a Vietnamese restaurant. (Daniel Fernández Photo Albums)

A wounded South Vietnamese soldier and an American soldier on a bike at the street market. (Daniel Fernández Photo Albums)

Buddhists temples in South Vietnam, 1965. (Daniel Fernández Photo Albums)

*American soldiers with Vietnamese men, women, and children.
(Daniel Fernández Photo Albums)*

*South Vietnamese civilians walking along road.
(Daniel Fernández Photo Albums)*

Foods available in a Vietnamese street market.
(Daniel Fernández Photo Albums)

Daniel and a friend in a Áo dài, a traditional Vietnamese dress.
(Daniel Fernández Photo Albums)

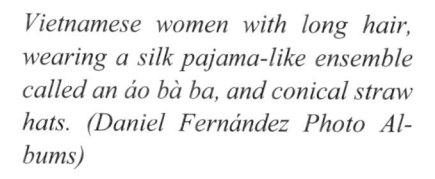

Vietnamese women with long hair, wearing a silk pajama-like ensemble called an áo bà ba, and conical straw hats. (Daniel Fernández Photo Albums)

During one of his missions as a helicopter shotgunner, Daniel was wounded in the leg as he helped three of his fellow soldiers under heavy enemy fire. For his wounds and heroism in action, Fernández received two Purple Heart citations accompanied by a Purple Heart medal in March 1965 and the Air Medal on May 25, 1965.

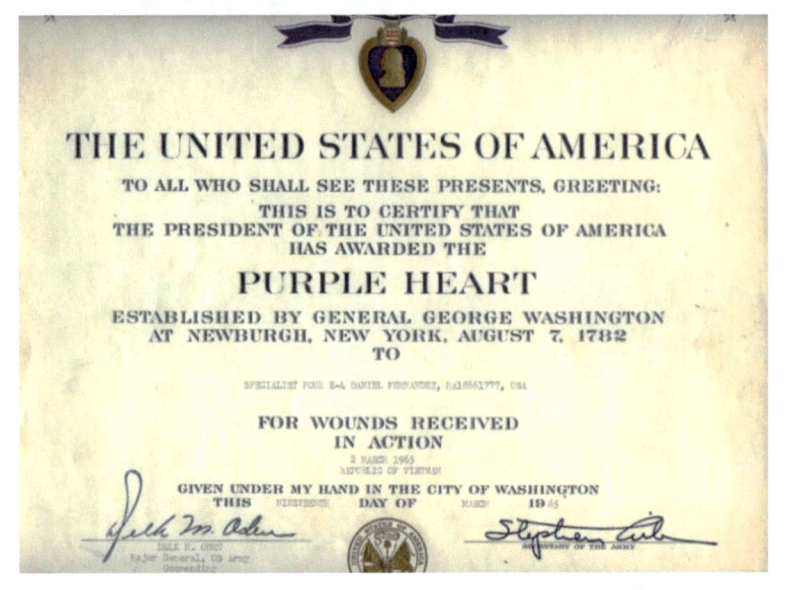

Purple Heart Citation awarded March 2, 1965, in the Republic of Vietnam. (Daniel Fernández Photo Albums)

Purple Heart Medal Daniel received. (Laurie Kastelic Collection)

Daniel D. Fernández: The Man Behind the Medal of Honor

CITATION

BY DIRECTION OF THE PRESIDENT
THE AIR MEDAL

IS PRESENTED TO

SPECIALIST FOUR E-4 DANIEL FERNANDEZ, RA18661777, USA

For distinguishing himself by meritorious achievement while participating in sustained aerial flight in support of combat ground forces of the Republic of Vietnam during the period

24 February 1965 to 17 March 1965

During this time he actively participated in more than twenty-five aerial missions over hostile territory in support of counterinsurgency operations. During all of these missions he displayed the highest order of air discipline and acted in accordance with the best traditions of the service. By his determination to accomplish his mission in spite of the hazards inherent in repeated aerial flights over hostile territory and by his outstanding degree of professionalism and devotion to duty, he has brought credit upon himself, his organization, and the military service.

Air Medal Citation for service in the Republic of Vietnam February 24, 1965, to March 1, 1965. (Fernández Collection)

United States Air Medal Daniel received. (Laurie Kastelic Collection)

46	Chapter 3: Training and First Deployment

When Shotgun IX's orders were extended 30 days in Vietnam, and the shotgunners were ordered into duty on the ground as a security force around the then-fledgling Da Nang airfield. They were then sent back to Hawaii where Danny recuperated before coming home to New Mexico on a 30-day furlough in June 1965.

Daniel in his western attire on horseback in Hawaii. (Fernández Collection)

Daniel posing with a young woman in Hawaii. (Daniel Fernández Photo Albums)

Before leaving Hawaii, Daniel had volunteered for another deployment to Vietnam. After visiting with his family, he returned to Schofield Barracks and underwent further combat training, including armored vehicle tactics and amphibious operations. By December 1965, the 25th Infantry Division was slated for service in Vietnam.

M113 Armored Personnel Carrier (APC) was designed to taxi troops within but was utilized as an armored fighting vehicle during the Vietnam Conflict. Vietnamese guerillas called the APCs "Green Dragons." (Daniel Fernández Photo Albums)

Daniel Fernández operating a .50 caliber M2 machine-gun on a M113 APC. (Daniel Fernández Photo Albums)

48   Chapter 3: Training and First Deployment

Training Maneuvers in M113 APCs in Hawaii. Each carrier's back dropped down to hold 11 soldiers. (Daniel Fernández Photo Albums)

Daniel D. Fernández: The Man Behind the Medal of Honor   49

USS Stone County *(LST-1141) on beach in Hawaii. (Daniel Fernández Photo Albums)*

Daniel trained on the USS Stone County *(LST-1141), a tank landing ship that the 25th Infantry Division used for amphibious training while in Hawaii. (Daniel Fernández Photo Albums)*

Chapter 3: Training and First Deployment

Deploying to Vietnam December 1965

Schofield Barracks, 1965. (Hazel Willis)

Troops, trucks, and APCs being transported on the USS Tioga County *(LST-1158) to Vietnam in December 1965. (Hazel Willis)*

Daniel D. Fernández: The Man Behind the Medal of Honor 51

25th Infantry Division, 1st Battalion, 2nd Brigade, 5th Infantry Mechanized preparing to leave Hawaii for Vietnam, December 1965. (Hazel Willis)

Chapter 3: Training and First Deployment

USS Tioga County *(LST-1158) arriving at Pearl Harbor, Oahu, with her crew manning the rails, 1965. (Navysource Naval History)*

Dan R. Shearin (left) leaving Hawaii on the USS Tioga County *(LST-1158). (Hazel Willis)*

Chapter 4: Second Deployment and Final Patrol

Daniel returned to Vietnam in December 1965. His division was taxied 6,000 miles across the Pacific on LST tank landing ships and troop ships arriving at the Port of Vung Tau in South Vietnam near Saigon. In Operation Blue Light, the 25th Infantry Division moved 4,000 jungle combat-trained troops and 9,000 tons of equipment in 25 days to the northwest sector of South Vietnam.

Daniel was assigned to the 1st Battalion of the 5th Mechanized Infantry within the 25th Infantry under the overall command of Major General Frederick C. Weyand.

The 5th Infantry Regiment had a distinguished history, dating back to 1808. They had acquired the nickname of "Bobcats" with the accompanying motto "I'll Try, Sir."

The 25th, nicknamed Tropic Lightning, with the accompanying motto of "Ready to Strike," was deployed to Cu Chi, an area about 20 miles northwest of Saigon and 18 miles from the Cambodian border.

As early as the 1880s, the Vietnamese who resisted the French occupation had dug a network of tunnels in the area to facilitate the resistance movement. By the time of the Vietnam

Shoulder insignia of the 25th Infantry Division. (Department of Defense)

5th Infantry shoulder insignia of "Bobcats." (Department of Defense)

Conflict, this network of tunnels had expanded into a complex network of nearly 250 miles.

The tunnels had multiple layers, constructed to minimize threats from bombing, gas attacks, and enemy infiltration. The outside entrances were carefully concealed and often booby-trapped.

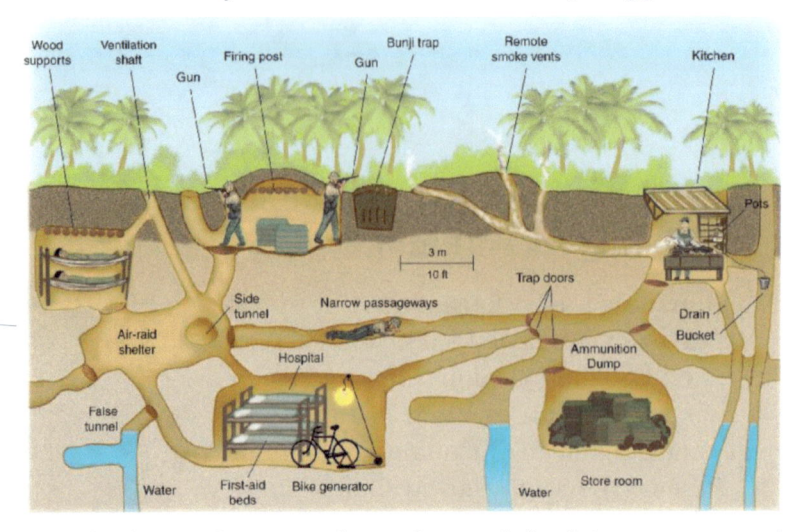

Multi-level Viet Cong tunnel complex used for living quarters and fighting. Based on displays at Vietnam Memorial at Ben Dinh near Saigon. (Open Journal of Soil Science)

Upon arriving in Vietnam, Daniel and the 1st Battalion Mechanized, 5th Infantry set up a staging area near Saigon University. The 1st Division's 3rd Brigade, under the command of Colonel William D. Brodbeck, was tasked with preparing the way for the 25th Infantry Division. Nearly every day the soldiers of the 3rd Brigade found ample signs of the Viet Cong's presence – booby traps, mines, weapons, ammunition, supplies, bunkers, and tunnel systems – but few enemy soldiers.

On January 24, 1966, certain that no large Viet Cong units were in the area, Colonel Brodbeck moved his brigade to Cu Chi proper to establish a perimeter around what was to become the 25th Division's new home. Almost immediately, his troops turned up a series of trenches and tunnels that seemed to permeate the site in every direction. On the morning of January 30, two companies from the 1st Battalion Mechanized, 5th Infantry passed through the lines of the 2nd Battalion 27th Infantry and attacked outward. With the 25th's 2nd Brigade largely in place, Brodbeck turned over control of Cu Chi to the brigade commander, Colonel Lynnwood M. Johnson, and returned to Lai Khe.

Daniel D. Fernández: The Man Behind the Medal of Honor 55

Personnel of the 2nd Brigade, 25th Infantry Division arrive at staging area near Bien Hoa, Vietnam, from their former base at Schofield Barracks, Hawaii. January 18, 1966. (SP5 Lawrence J. Sullivan, US Army)

Troops of the 2nd Brigade, 25th Infantry Division set up tents at staging area near Bien Hoa, Vietnam, after arrival from Hawaii. (SP5 Lawrence J. Sullivan, US Army)

The 2nd Battalion 27th Infantry with Company B, 65th Engineers followed, sweeping and clearing the area and destroying tunnel complexes that were located. During the five-day operation 20 tunnel complexes, some as long as a half a mile, were located and destroyed.

Colonel Johnson's troops spent much of the next few months trying to clear their base camp of the maze of tunnels that lay beneath it. The Viet Cong would appear in the middle of the camp through hidden access points to the tunnel complex. It was only after using specially trained soldiers known as "tunnel rats" that the camp became immune to internal attack. Although the tunnel access points had been removed, Viet Cong sympathizers managed to infiltrate the camp and provide advance notice to their comrades of impending attacks.

As a part of their mission to defeat the Viet Cong insurgency in the area, the various units at Cu Chi sent out daily patrols to identify enemy strongpoints. In addition, these patrols served to protect the perimeter of the sprawling Cu Chi complex. Between his arrival at Cu Chi in January and his final patrol on February 18, Daniel would have been involved in almost daily patrols and counter-insurgency operations in the vicinity of the base camp. The citation for the Valorous Unit Award for the 2nd Brigade of the 25th Infantry Division said, in part, that the Brigade conducted 11 major operations with near-constant combat against the Viet Cong over a 66-day period from January 30 to April 5.

Norman "Butch" Petit was one of the first "tunnel rats" to run the hole in Cu Chi for Bravo Company 1st BN (M), 5th infantry. He was awarded the Soldier's Medal for pulling wounded soldiers from a burning APC in December 1966. (Butch Petit)

SGT Ronald H. Payne exiting a tunnel in search of Viet Cong soldiers dubbed "Magic Charlie" by U.S. forces. (Wikimedia Commons)

A "Tunnel Rat" soldier with gas mask and gun. (The Chive)

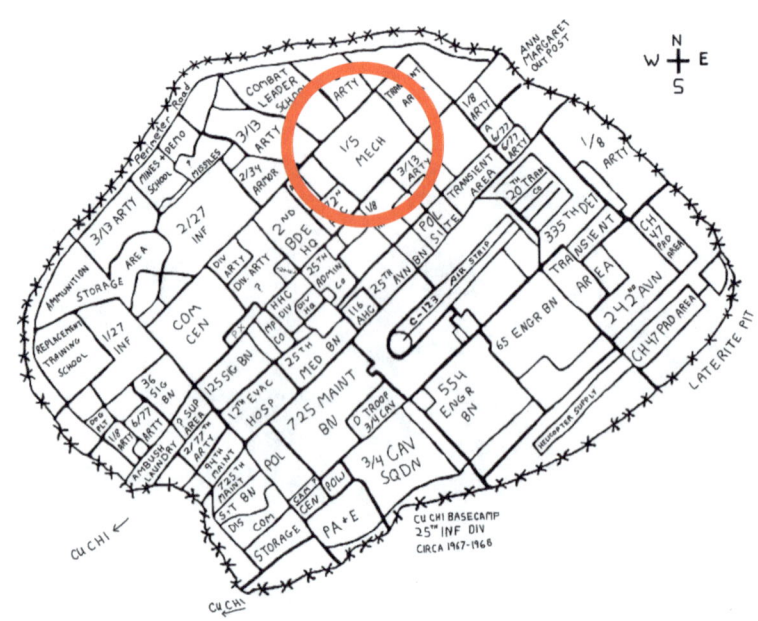

58　　　　　　　　　　Chapter 4: Second Deployment and Final Patrol

Map of 25th Infantry Division Base Camp at Cu Chi, 1967-68. Daniel's unit 1ˢᵗ Battalion, 5ᵗʰ Mechanic Infantry circled. (Wolfhound Alumni)

Aerial view of 25th Infantry Division Headquarters at Base Camp near Cu Chi, March 27, 1966. (SSG Howard C. Breedlove, US Army)

Camp at Cu Chi. (Larry Artiaga Collection)

Water purification truck at Cu Chi. (Larry Artiaga Collection)

Makeshift beds with mosquito netting. (Larry Artiaga Collection)

60		Chapter 4: Second Deployment and Final Patrol

Daniel at his writing desk in Vietnam. (Fernández Collection)

Photos of Daniel in Vietnam. (Top: Unknown, Bottom: Albuquerque Tribune*)*

On January 31, 1966, one day after Colonel Brodbeck turned the area over to Colonel Johnson, 1st Battalion, 5th Infantry, 25th Infantry Division experienced its first losses. The Battalion had several patrols out along the Soui Ben River near the Filhol Rubber Plantation that became known by U.S. soldiers as the Gateway to the Hồ Bò Woods[2].

SGT Dan R. Shearin had just led a patrol into the area and upon their return the team decided to break and eat some C Rations while they waited for the other patrols. As they sat to eat, SGT Shearin and SP4 Armando Tesillo accidently set off a Viet Cong booby trap along the bank of the river, killing them instantly.

SGT. Dan R. Shearin,
Jan. 24, 1946 – Jan. 31, 1966.
(Hazel Willis)

SP4 Armando Tesillo
Sept. 25, 1946 – Jan. 31, 1966
(Fold3.com)

Shearin, a North Carolina native, had just turned 20 years old on January 24 and Tesillo, a Los Angeles native, was 19. Daniel, and the two men had served together in Vietnam as door gunners in 1965 in Shotgun IX. The men saw a lot of combat on their first tour of duty. They brought a lot of experience which was vital to those that deployed

---

[2] The Hồ Bò Woods are located in Bình Dương province 20 km north of Củ Chi, 4 km to the west of the Iron Triangle and the Saigon River and 56 km northwest of Saigon. The woods consisted of rubber plantations, sparse dense woods, and open rice paddies. The Viet Cong used the woods as a base camp during the Vietnam Conflict.

from the Battalion from Hawaii to Cu Chi. Veterans on the 1st Bn (M), 5th Infantry, 25th Infantry Division Bobcat website said:

> The event emotionally jolted some of the men of the company and impressed upon them that this was the real thing. These were people they knew, if only by sight. One moment laughing, talking, breathing, living, and the next moment they are gone. How thin and delicate the thread between life and death is. It was a lesson soon to be indelibly implanted in the consciousness of the soldiers of the battalion. Some wondered who would be next.

Dan Shearin and Armando Tesillo's deaths were the first of 172 deaths 1st Bn (M), 5th Infantry, Class of 1966 would experience that year. On February 11, 1966, one man from Company B was shot while examining the kill zone after a nighttime ambush was tripped. No one used a flashlight after that. Fourteen Bobcats would be killed in action in the month of February 1966, including SP4 Daniel D. Fernández and SGT Joseph T. Benton.

The following account of Daniel Fernández' final patrol is taken from several sources, specifically the official report history of the 1st Battalion of the 5th Infantry Division, news articles, and the November 2000 edition of the *Tropic Lightning News*.

> On February 18, 1966, the First Platoon of Company C began a raiding mission at 0100 hours. All 18 members of the raiding party were volunteers. To no one's surprise, SP4 Daniel Fernández was one of the men who volunteered. Second Lieutenant Joseph D'Orso pointed out that Daniel had participated in a patrol in the same area on the previous day and had less than four hours sleep before the platoon set out on the next patrol.
>
> They crossed their line of departure (LD) at about 0115 hours and the Ben Muong [creek] at 0130 hours. After crossing the stream, they moved through dry rice paddies for about 900 meters, stopping some 25 meters short of the wood line, where they set up security.
>
> Fernández and the squad leader went into the wood line on a brief reconnaissance of the area. At about 0235 hours, the two men returned, and the party moved forward into a rubber plantation for about 300 meters. The raiding party set up a perimeter and Fernández and the squad leader went out on another reconnaissance of the area. When they returned to the raiding party, they stated that there

was nothing to raid. The party then moved back about 100 meters and set up in a line perimeter near a graveyard where they were going to wait until moving back to the base camp at 0900 hours.

At about 0700 hours, SP4 Joseph T. Benton of Hetford, North Carolina, spotted seven Viet Cong soldiers on the left front behind a burned-out hut. Benton opened fire with his M-60 machine gun, killing three of the Viet Cong. He was reaching for a hand grenade when he was shot and killed by a Viet Cong sniper. Immediately the entire patrol came under intense enemy small arms, automatic weapons, and grenade fire. The left flank was pushed back about 20 meters.

Fernández crawled to one side of the hut and SP4 James P. McKeown of Willingsboro, New Jersey, moved into place on the other side of the hut. Behind the hut PFC David R. Masingale of Fresno, California, the medic, moved forward in an attempt to rescue Benton. Three more men came out to help. Benton was picked up and the group started to move back when one of the rescuers, SGT Ray E. Sue, was shot in the left leg, knocking him to the ground. The other men, including SP4 George E. Snodgrass of Pompton Lakes, New Jersey, hit the ground and returned fire.

PFC Masingale moved to treat SGT Sue while the other men riddled the bushes with small arms fire. Grenades and small arms and automatic weapons fire from the Viet Cong was intense. At that instant a grenade fired from a rifle by one of the guerrillas landed by Fernández's leg. He accidentally kicked the grenade when he attempted to move away, and it rolled nearer to Benton and Sue.

Without hesitation, so quickly that PFC Masingale was sure he didn't have time to consider the consequences of his action, Fernández shouted, "move out, you people", dove onto the grenade, and smothered the blast with his body, saving the lives of those around him.

Specialist 4th class George Snodgrass rigged a litter out of bamboo poles and three shirts, so that Daniel could be moved to a more open area where a MedEvac helicopter could land. As they waited for the helicopter, Masingale realized that Daniel's wounds were life-threatening. Besides wounds inflicted by the exploding grenade, Daniel had been hit by a stray enemy bullet. When the grenade exploded, its fragments ripped into his groin, his abdomen, and his right leg. Fernández was bleeding badly. "Hang on, Buddy," the medic told Fernández. In great pain, Daniel replied, "I'm going to hang on," but added, "I never believed it would hurt so much."

Snodgrass, a devout Roman Catholic who frequently attended Mass with Daniel, urged him to make an act of contrition since there was no priest to hear his confession. Daniel said, "I will."

Artillery and mortar fire was called in on the enemy positions around the patrol. Tactical air strikes with napalm were also summoned. The remainder of the 1st Platoon of Company C made their way out to the patrol. At that time the raiding party moved back to the rice paddy area at the edge of the woods. Fernández, gravely injured but still conscious, was laid on the ground, and SGT Sue was laid next to him. Sue held Fernández's hand and talked with him. Fernández told him he was hurting. The Dustoff[3] arrived and, with the aircraft's blades still rotating, medics rushed Daniel on board.

Sergeant Ruben Perkins, nicknamed "SGT Rock" by Fernández, had spoken to "Old Dan" (Perkins' nickname for Daniel, since, at the age of 21, he was older than most of the 18 and 19-year-olds in the unit) before the helicopter took off. Still in good spirits despite his serious condition, Daniel kiddingly asked Perkins, "Who's going to take care of you now?" They had been working together since Daniel joined the unit. Sadly, SGT Perkins was to die in combat within weeks.

Daniel was airlifted back to his brigade's hospital. SGT Sue was placed next to Fernández at the hospital. The next time he looked, Fernández had been moved. Sue asked the doctor about Fernández, but the doctor would not tell him anything.

The Army doctors on duty fought for two hours to stop Daniel's internal bleeding. Their hard work was to no avail. The damage to Daniel's body was too great. He died that day in the Army hospital.

The 1st Platoon of Company C returned to the base camp perimeter by 0935 hours. According to the after-action report, the Viet Cong lost as many as 22 men during the fight.

---

[3] Dustoffs were the helicopter ambulances, usually the Bell UH-1 Iroquois nicknamed Hueys, that were used for MedEvac purposes. The name originated with the call sign for the 57th Medical detachment.

A typical MedEvac operation. During those very risky moments, the medic and crew chief would jump into the line of fire and load casualities into the helicopter with the assistance of nearby soldiers. (Flickr: Photo by Bill Hall, 1966/Michael Ochs Archive)

There were NBC news correspondents with Daniel's company capturing the action to bring the events of the war to viewers back home in the United States. The time-worn video the news crew captured of Daniel's platoon coming back from the battle is very telling.

Smoke is billowing up from behind the trees. You can hear mortars exploding. Then slowly, a single line of solemn looking men weaves their way across a dry, beaten down field of rice. They are wearing sweat-stained army t-shirts and fatigues. They are dirty and some have their faces blackened for concealment during the night patrol. Their helmets are camouflaged with leaves while others wear caps or nothing at all to protect their heads. They are carrying rifles over their shoulders and by their sides. The men in the base camp can see they are weary and carry a burden of sadness and a burden of relief. They have survived.

One man can be heard saying: "Give them room. Give them room." He continues:

> They were throwing grenades like mortars. They were coming in through the air. Right down on us. They came right down and hit Fernández in the mid-section. When they went after Fernández, we caught it from the side. We started getting it

from three machine guns from one side to other side to the middle. They were hitting us all around with machine guns and the grenades were coming in from all over the place. I'll tell you, there is a company out there…We got our share.

One of the commanders disclosed:

They were the best I had. I thought we had the most firepower last night. We sent them out there and had a well-planned ambush setup. They had so many darn tunnels out there and trench networks. These sneaky little devils would just come up; in on you from all sides, center, and use automatic weapons on you but from what we had out there, we had some of the finest people in the United States Army. They did their best and a couple of them deserve Congressional Medals of Honor.

The following are snapshots from the NBC News video:

Soldiers waiting for Daniel's unit to return from night patrol. (NBC News)

A machine gunner watching 1st Platoon walk across field. (NBC News)

1st Platoon soldiers emerging from jungle. (NBC News)

1st platoon soldiers, some with blackened face for concealment during the night patrol. (NBC News)

Members of the 1st Platoon returning from Daniel's fatal patrol. (NBC News)

68　Chapter 4: Second Deployment and Final Patrol

1st Platoon returning from Daniel's fatal patrol. (NBC News)

Men from 5th Infantry debriefing after the incident. (NBC News)

SP4 James P. McKeown, one of the men Daniel saved. (NBC News)

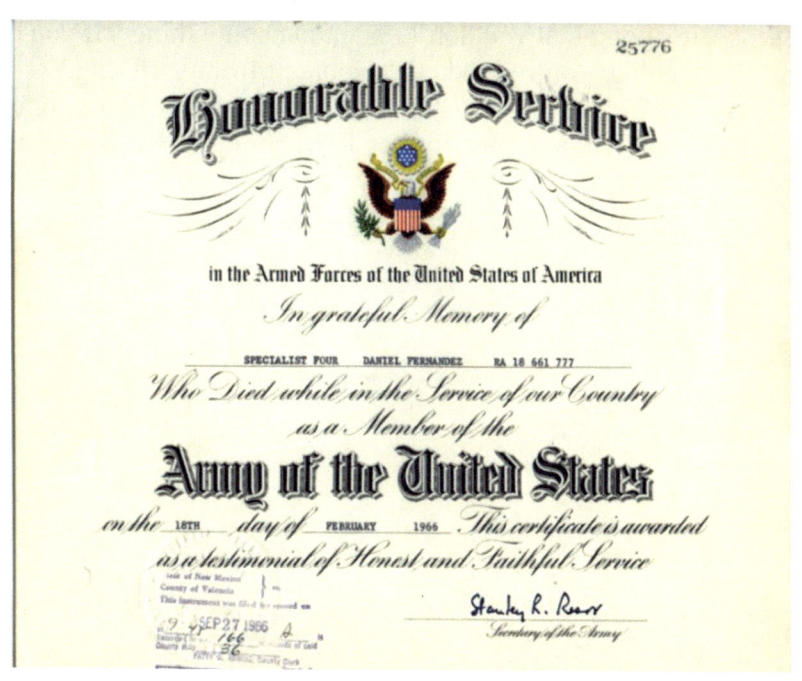

Daniel D. Fernández: The Man Behind the Medal of Honor

Recognition of Honorable Service at the time of Daniel's death. In April of 1966, Daniel's parents accepted another Purple Heart on behalf of their son for wounds received on February 18, 1966. Rear Admiral Ralph C. Johnson, retiring commander of Sandia Base, presented the medal and citation. The award was Daniel's second Purple Heart Medal and third Purple Heart Citation. He received two Purple Hearts for the injuries he sustained in March 1965 as a helicopter gunner. (Fernández Collection)

During World War II, Korea, and Vietnam, the military employed Western Union to communicate official notifications to the families of service members. This included informing loved ones that their relative had been wounded, missing in action, taken prisoner, or killed in action. With limits on how many words and punctuation in each telegram, the military adopted a distinct standardized phrasing to convey the news; hence the common chilling phrase "I regret to inform you…"

In the Fernández family's case, José was notified first. He went to Laura to share the news. Cecilia Otero Castillo remembers the day and Laura's absolute devastation. Laura was alone working as an appraiser in her office at the Valencia County Courthouse when she looked up to see José and some unknown man standing in the door.

"I knew, I just knew what happened," Laura said. "After that, everything was a blur and sometimes I couldn't remember who visited me.

In a sense, it's good that you don't remember some things. If you erase them, you feel like you're protecting yourself from suffering."

"*Es un valle de lagrimas* (it is a valley of tears) when you lose a son," Laura told Angie Lopez for her 1990 book, *Blessed are the Soldiers*.

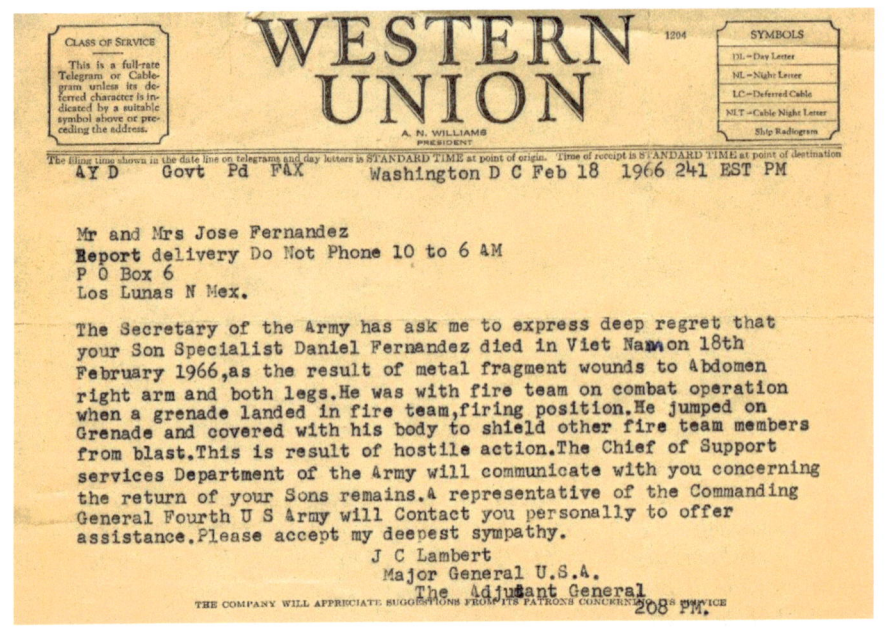

```
CLASS OF SERVICE        WESTERN           1204   SYMBOLS
This is a full-rate                              DL—Day Letter
Telegram or Cable-       UNION                   NL—Night Letter
gram unless its de-                              LC—Deferred Cable
ferred character is in-                          NLT—Cable Night Letter
dicated by a suitable   A. N. WILLIAMS           Ship Radiogram
symbol above or pre-     PRESIDENT
ceding the address.
The filing time shown in the date line on telegrams and day letters is STANDARD TIME at point of origin. Time of receipt is STANDARD TIME at point of destination
    AY D    Govt  Pd  FAX   Washington D C Feb 18  1966  241 EST PM

         Mr and Mrs Jose Fernandez
         Report delivery Do Not Phone 10 to 6 AM
         P O Box 6
         Los Lunas N Mex.

         The Secretary of the Army has ask me to express deep regret that
         your Son Specialist Daniel Fernandez died in Viet Nam on 18th
         February 1966, as the result of metal fragment wounds to Abdomen
         right arm and both legs. He was with fire team on combat operation
         when a grenade landed in fire team, firing position. He jumped on
         Grenade and covered with his body to shield other fire team members
         from blast. This is result of hostile action. The Chief of Support
         services Department of the Army will communicate with you concerning
         the return of your Sons remains. A representative of the Commanding
         General Fourth U S Army will Contact you personally to offer
         assistance. Please accept my deepest sympathy.
                                J C Lambert
                                Major General U.S.A.
                                The Adjutant General
         THE COMPANY WILL APPRECIATE SUGGESTIONS FROM ITS PATRONS CONCERNING ITS SERVICE
                                                              208 PM.
```

Telegram José and Laura received from Major General J.C. Lambert, the Adjutant General, informing them that Daniel died on February 18, 1966. (Fernández Collection)

There was one person that was with the Fernández family from the time he heard the news of Daniel's death. Father Francis Schuler was there every day giving the Fernández family comfort and friendship. He was a tremendous help in their time of need.

Chapter 5: Coming Home

Daniel's body was flown to Travis Air Force Base in California and transported to the train depot in Richmond, California. The Army-issue, burnished steel coffin was draped with an American flag and loaded aboard Santa Fe train Number 2, the eastbound San Francisco Chief. It was escorted by Specialist First Class Anthony Aguirre from Socorro, New Mexico, and Lance Corporal Lee Pullins from Albany, Georgia. Both men had asked if they could escort the casket on its journey to Belen. The train left Richmond at noon on February 24, 1966.

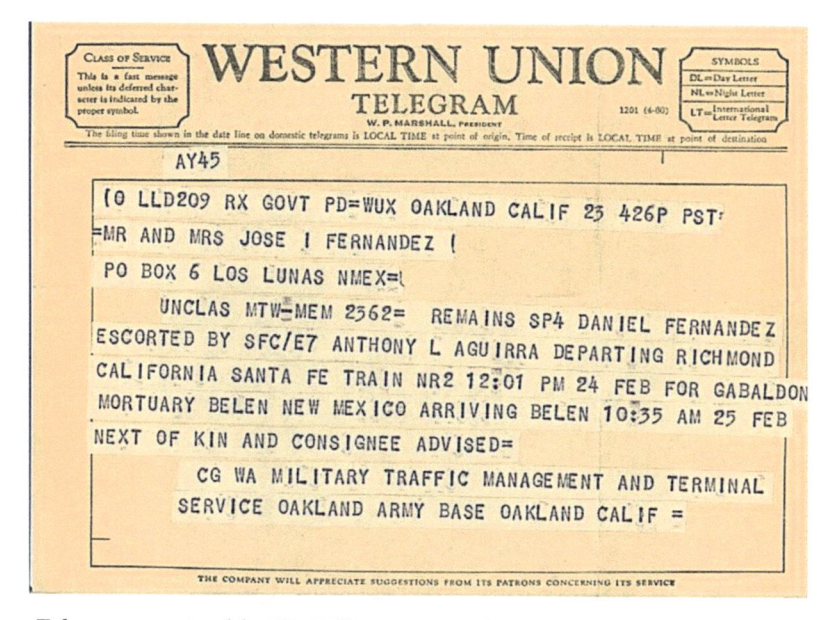

Telegram received by Daniel's parents informing them when and where his remains would arrive. (Fernández Collection)

Chapter 5: Coming Home

Daniel's body arrived in Belen by train and was escorted by the U.S. Army Element from Sandia Base. (White House Memorial Album)

On February 26, Daniel's body lay in state until 8 PM, at which time a Rosary was recited at the San Clemente Catholic Church in Los Lunas.

Father Francis Schuler performing a High Requiem Mass. (Albuquerque Journal)

At 8:30 AM on February 27, a crowd of approximately 1,300 gathered at the Los Lunas High School on Luna Avenue, filling the high school's gymnasium. The gym had been hastily prepared to handle the huge crowd, which was much too large for San Clemente Church where the funeral had originally been scheduled. Father Francis Schuler, pastor of San Clemente, celebrated the High Requiem Mass.

In his homily, Father Schuler recalled the oft-quoted verse from the Gospel of John, chapter 15, verse 13: "Greater love hath no man than this, that a man lay down his life for his friends." He noted:

> Daniel Fernández went to meet his God out of love for his country, for his fellow soldiers, for us. For our God and for our great America Daniel Fernández lived and fulfilled the high ideals once urged by President Kennedy: "Ask not what your country can do for you, but rather what you can do for your country."
>
> Daniel had the kind of patriotism that rises above creed and class…. We have gratitude for the supreme sacrifice this young man made…. He has taught us the true meaning of patriotism.
>
> We can never wipe out the terrible cost of war in blood and tears….

State Police Chief John Bradford assigned Officer Manuel Otero Jr. of Los Lunas to escort the 90-car procession to the Santa Fe National Cemetery for the burial service.

As the procession reached the National Cemetery and moved slowly to the gravesite, nearly 2,000 people had been waiting in the cold damp afternoon for the cars to arrive from Los Lunas. They were joined by Major General John P. Jolly, New Mexico's adjutant general, by Brigadier General Burton R. Brown, the senior military officer at Sandia Base, and by Mrs. Joseph Montoya, wife of New Mexico senator Joseph P. Montoya.

After Father Schuler's remarks, Daniel's two brothers, James and Peter, and his sister, Rita, stepped forward to place white carnations on the casket. Then his parents placed a spray of red, white, and blue carnations upon his casket, followed by a 21-gun salute. As the honor guard took the American flag that was ceremoniously draped over Daniel's coffin and folded it, military Taps, played by a bugler, rang out across the hillside. Then, in a token of respect and appreciation, the flag was presented to his parents for Daniel's service.

Major General John P. Jolly, Brigadier General Burton R. Brown, José, Laura and Rita Fernández after services at Los Lunas High School. (White House Memorial Album)

Pall bearers were members of the U.S. Army Element from Sandia Base. (White House Memorial Album)

Military escort after services at the Los Lunas High School Gymnasium. (White House Memorial Album)

76   Chapter 5: Coming Home

The Fernández family knelt quietly during services for Daniel. Mrs. Fernández with her face covered in a black mantilla, knelt quietly with her husband and children, Rita, James, and Peter.
(Valencia County News)

Funeral cortege from Los Lunas to Santa Fe.
    (Albuquerque Journal)

*Eight-gun salute at Santa Fe National Cemetery.
(White House Memorial Album)*

Mourners of all ages attended the services. (White House Memorial Album)

78   Chapter 5: Coming Home

Father Schuler recites graveside prayer. (White House Memorial Album)

U.S. flag being ceremoniously folded by members of the Sandia Base First Element. (White House Memorial Album)

José and Laura Fernández receiving the United States flag honoring their son, Sp4 Daniel D. Fernández.
(White House Memorial Album)

Chapter 5: Coming Home

Daniel was laid to rest in a choice location at the Santa Fe National Cemetery—at the highest point in the cemetery, atop a hill under a 30-foot-tall pine tree. The nearest grave was that of SSGT Dennis Patterson, a member of the 34th Infantry in World War II. Forty feet away was the grave of Major General Patrick Hurley, famed military leader, former United States Secretary of War (1929-1933), and statesman. Daniel's original headstone, installed in 1966, was replaced with the Medal of Honor headstone after he received the Medal in 1967.

Flag at half-staff over Santa Fe National Cemetery in honor of Daniel Fernández in 1966.
(Albuquerque Journal)

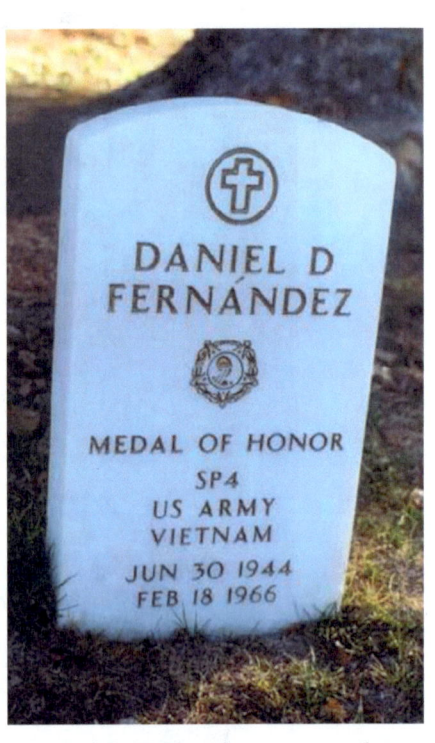

Daniel D. Fernandez grave marker.
(Laurie Kastelic Collection)

Daniel shares the hallowed ground of the Santa Fe National Cemetery with eight other New Mexico Medal of Honor recipients (see Appendix 3).

Later that day, back in Valencia County, several dignitaries paid homage to this hero of Homeric bearing. U.S. Senator Joseph Montoya spoke to an audience of about 200 in Belen after visiting with the Fernández family in Los Lunas. The Senator described the Fernández home as "sad but proud." Montoya declared, "The whole country feels our loss in New Mexico for this boy who died to save his comrades." The lawmaker concluded his remarks with the simple words, "I salute Daniel Fernández."

In Los Lunas, Mayor Howard Simpson declared Saturday, February 26, as Daniel Fernández Day. He requested that every home and business in town fly an American flag at half-staff in honor of the local hero.

In the aftermath of the funeral and burial, tributes arrived from all over the world. As many as 100 letters per day came from Korea, Ethiopia, and across the United States, including Puerto Rico. Senator Joseph Montoya loaned the use of his office and staff in Santa Fe, but it still took a week to reply to every piece of mail.

Others offered their assistance as well. Valliant Printing Company donated the stationary and the Lions and Rotary Clubs donated the postage. "Everyone was so good to us," Laura said. "I hope no one will feel bad if I've left them out."

Those that knew Daniel offered their memories. Becky Richins, who Daniel hoped to marry one day, recalled, "This is how I remember him—always willing to do something for someone else." Robert Schwartz, a junior high school student in Brooklyn, New York, wrote to Laura and José, "Your son will always be my hero. He was fighting for my freedom, and I am proud of him." The parents of Sergeant Ray Sue, one of the men whose life Daniel saved, said, "How humble we feel." High school counselor Pete Pino noted Daniel's "generosity" which gave him a sense of individual rather than group leadership. His childhood friend Frank Gurulé said that Daniel possessed an "inner strength" which was hard to describe but which was obvious to everyone who knew him.

Daniel's former comrades in arms recalled, "he was a friend to everyone, generous with his money when others had run out, a likeable guy." A fellow platoon member, Second Lieutenant Joseph D'Orso, said, "Everyone was Fernández's friend. He was always volunteering. He was in the same spot [where he was killed] the night before and in the platoon. He was ready to do anything. And he was always cheerful, even when he came out of [a] swamp covered with leeches."

Daniel's father made a salient comment when interviewed about his son by a *Valencia County News* reporter: "I would say he evidently had a love for his fellow men." He also mentioned, "Daniel was a man who didn't have any hatred. He had lots of friends among the South Vietnamese. He also had respect for the Viet Cong. He considered them very brave people, very tough." José added that Daniel's combination of generosity and courage was a "fatal mixture." He said, "Those qualities mean almost certain death in a situation like Vietnam. A man tries to raise a man. I didn't want a coward [but] I didn't want a dead hero."

He finished the interview by saying, "When I heard it [that Daniel had died], I just couldn't believe it. It's so unreal. I can say that I feel humble to know that my son did such a thing."

Medal of Honor Presentation
by
Lyndon Baines Johnson
President of the United States of America
to
Specialist Four Daniel Fernandez
(Posthumously)
United States Army

AT THE WHITE HOUSE
WASHINGTON, D. C.
on
THURSDAY, 6 APRIL 1967
at
1300 HOURS

Program for the Medal of Honor Presentation (Larry Artiaga Collection)

Chapter 6: Medal of Honor

On February 21, 1966, Daniel's company commander, Captain James G. Andress, nominated him for the Medal of Honor, and the nomination was endorsed by the division hierarchy. To those who fervently believed that the award should be bestowed, it seemed that the wheels of the Army bureaucracy ground on forever. The recommendation underwent a thorough review, going through a dozen headquarters in the Army chain of command before it reached the desk of Secretary of Defense Robert McNamara in Washington, D.C. McNamara approved the recommendation in October 1966 and sent it on to the Joint Chiefs of Staff. The Joint Chiefs concurred that Daniel deserved our nation's highest military honor.

Finally, after intervention by various New Mexicans, led by Representative E. S. "Johnny" Walker and Senator Clinton P. Anderson, the award ceremony was scheduled for 1 PM at the White House on April 6, 1967.

Daniel's family, including José, Laura, Rita, Peter, and James, together with Father Schuler, Elsie Ortiz (a cousin), and SGT Anthony Aguirre, traveled to Washington for the ceremony.

Presentation speech by President Lyndon B. Johnson

> Mr. and Mrs. Fernández, distinguished Members of Congress, Secretary Vance, Secretary Resor, General Abrams, General Johnson, ladies and gentlemen:
>
> We have come here to the Rose Garden today to speak of a brave young American who gave his life for us in in Vietnam.

Specialist Four Daniel Fernández earned his country's highest honor by a classic act of courage and self-sacrifice. He threw himself on a live grenade that had been fired among his comrades. By that act, he saved the lives of four other Americans. Two of them are with us today.

Daniel Fernández died before he was 22 years old. He was not yet born when other Americans crossed the Pacific in World War II. He was not yet in school when others went again to fight for freedom in Korea. Yet today, and forever, he is joined in the legion of American heroes.

The land in Asia where he gave his young life is half a world away from his home in Los Lunas, New Mexico. Yet he did not see much of the world. He went to school in Albuquerque and Los Lunas. He enlisted in the Army after high school, took basic training in Louisiana, [and] served in Hawaii for a time. Then he went to Vietnam, to a deserted hamlet northwest of Saigon, and finally to his fatal encounter with the Viet Cong grenade.

Daniel Fernández died on February 18, 1966. He died less than three weeks after we, in our ceaseless search for peace, had made our longest pause in the bombing of the North.

The question that haunts me today should concern every American. It is this: Was that grenade on one of the trucks or on one of the trains, or on one of the sampans that we let pass unmolested during that long 37-day pause?

If it was, Daniel Fernández died as more than a hero of battle, he died as a martyr in the search for peace.

And those who today are urging an unconditional permanent cessation of bombing should ask themselves: "What are the consequences?" It is one thing to talk abstractly of peace and war. It is something quite different to think of a young man named Daniel Fernández who will dream no more.

Mr. and Mrs. Fernández, in the name of the Congress, I pass to you the Medal of Honor of the United States, won so deservedly by the hero who was your son.

I give you this, our Country's greatest honor.

It is poor compensation for your loss. But be assured that the death of your son will have meaning. For I give you my solemn pledge that our country will persist – and will prevail – in the cause for which your boy died.

Mr. Stanley R. Resor, Secretary of the Army, will now read the citation.

Secretary of the Army Stanley Rogers Resor
(Department of Defense)

The President of the United States of America, authorized by Act of Congress, March 3, 1863, has awarded in the name of The Congress the Medal of Honor, posthumously, to

Specialist Four Daniel D. Fernández, United States Army

for conspicuous gallantry and intrepidity at the risk of his life above and beyond the call of duty:

Specialist Four Daniel Fernández distinguished himself by gallantry and intrepidity at the risk of his life above and beyond the call of duty on February 18, 1966 while serving as a member of an eighteen-man patrol engaged in a vicious battle with the Viet Cong in the vicinity of Cu Chi, Hau Nghia Province, Republic of Vietnam.

Specialist Fernández demonstrated indomitable courage when the small patrol was ambushed by a Viet Cong rifle company and driven back by the intense enemy automatic weapons fire before it could evacuate an American soldier who was down in the initial attack. Specialist Fernández, and three comrades immediately fought their way through devastating gun fire and exploding grenades to reach the fallen soldier. After the volunteers reached their fallen comrade and attempted to return to their defensive positions, a United States Army sergeant was struck in the knee by .50 caliber machine gun fire. Specialist Fernández rallied the left flank of his patrol, went to assist in the recovery of the wounded sergeant and, while first aid was being administered to the wounded man, an enemy rifle grenade landed in the midst of the group. Realizing there was no time for the wounded sergeant or the other men to gain protection from the grenade blast, Specialist Fernández threw himself on the grenade as it exploded, saving the lives of the four men at the sacrifice of his life.

Specialist Fernández's profound concern for his fellow soldiers, his conspicuous gallantry, and his intrepidity at the risk of his life above and beyond the call of duty are in the highest traditions of the United States Army and reflect great credit upon himself and the armed forces of his country.

Daniel was the first man from the 25th Infantry Division to receive the honor during the Vietnam Conflict. Daniel was the first of 22 Hispanic Americans to be awarded the Medal of Honor in the Vietnam Conflict[4]. It was the ninth such medal personally presented by the President, the 14th of the war in Vietnam, and the 2,419th in history.

The recommendation and subsequent award of the Medal of Honor to Daniel came in the midst of the Civil Rights movement and stood as a milestone for Hispanic Americans. In the past, Hispanic, Black, Jewish, and other Americans were overlooked for military commendations in favor of their white counterparts. This prejudice extended to their hometowns where even burials were segregated.

---

[4] Daniel held that distinction until President George W. Bush posthumously awarded the Medal of Honor to Captain Humbert Roque "Rocky" Versace in 2002.

One such instance was World War II veteran Private Felix Z. Longoria of Three Rivers, Texas. Longoria was born and raised in Texas. He enlisted in the Army in November 1944 and was assigned to a regiment fighting in the Philippines. On June 16, 1945, fifteen days after landing in Luzon, Private Longoria was killed in an ambush that later earned him a Purple Heart and other medals.

It took four years for his remains to be recovered and repatriated. According to his family, a funeral home in the small South Texas town refused to hold his funeral services in the mortuary because he was Hispanic. The funeral home offered to bury him in the "Mexican section" of the cemetery, segregated by barbed wire but they couldn't hold services in the mortuary because "The whites wouldn't like it."

The American GI Forum (AGIF), a Hispanic veterans and civil rights organization founded in 1948, brought national attention to the case, and the then-freshman Texas Senator Lyndon B. Johnson obtained authorization for Longoria's remains to be buried at Arlington National Cemetery in Arlington, Virginia. On February 16, 1949, Longoria was laid to rest with full military honors, with Senator and Lady Bird Johnson in attendance. People gathered at his grave annually for decades, and the "Felix Longoria Affair" played a significant role in catalyzing Mexican American political consciousness and activism that helped lay the grounds for Daniel's recognition.

In the months after Daniel's death, the AGIF branch in Santa Fe assisted Daniel's parents in setting up a Daniel Fernández Education Scholarship, which eventually led to the formation of the Daniel Fernández Chapter of the AGIF.

Fernández family arriving at National Airport, Washington, D.C., on April 5, 1967. (White House Memorial Album)

José and Laura Fernández. (White House Memorial Album)

Reverend Schuler, Elsie Ortiz (cousin), Sergeant Anthony Aguirre, José, Laura, Rita, James, and Peter Fernández at the Tomb of the Unknown Soldier. (White House Memorial Album)

Chapter 6: Medal of Honor

President Lyndon B. Johnson greets Daniel's father, José, at the Medal of Honor ceremony. (White House Memorial Album.)

Medal of Honor given to the Fernández Family. (Fernández Collection)

The President of the United States of America, authorized by Act of Congress, March 3, 1863, has awarded in the name of The Congress the Medal of Honor, posthumously, to

Specialist Four Daniel Fernandez, United States Army

for conspicuous gallantry and intrepidity in action at the risk of his life above and beyond the call of duty:

Specialist Four Daniel Fernandez distinguished himself by gallantry and intrepidity at the risk of his life above and beyond the call of duty on February 18, 1966 while serving as a member of an eighteen-man patrol engaged in a vicious battle with the Viet Cong in the vicinity of Cu Chi, Hau Nghia Province, Republic of Vietnam. Specialist Fernandez demonstrated indomitable courage when the small patrol was ambushed by a Viet Cong rifle company and driven back by the intense enemy automatic weapons fire before it could evacuate an American soldier who was struck down in the initial attack. Specialist Fernandez and three comrades immediately fought their way through devastating gun fire and exploding grenades to reach the fallen soldier. After the volunteers reached their fallen comrade and attempted to return to their defensive positions, a United States Army sergeant was struck in the knee by .50 caliber machine gun fire. Specialist Fernandez rallied the left flank of his patrol, went to assist in the recovery of the wounded sergeant and, while first aid was being administered to the wounded man, an enemy rifle grenade landed in the midst of their group. Realizing there was no time for the wounded sergeant or the other men to gain protection from the grenade blast, Specialist Fernandez threw himself on the grenade as it exploded, saving the lives of the four men at the sacrifice of his own. Specialist Fernandez' profound concern for his fellow soldiers, his conspicuous gallantry, and his intrepidity at the risk of his life above and beyond the call of duty are in the highest traditions of the United States Army and reflect great credit upon himself and the armed forces of his country.

Daniel's Medal of Honor Citation. (Larry Artiaga Collection)

94   Chapter 6: Medal of Honor

First Lady, Lady Bird Johnson, greets José and Laura Fernández at the Medal of Honor ceremony. (White House Memorial Album)

President Lyndon B. Johnson presenting the citation and Medal of Honor to Daniel's father. (White House Memorial Album)

James, Rita, and Peter Fernández during the ceremony. (White House Memorial Album)

President Lyndon B. Johnson speaking with Daniel's parents José and Laura Fernández in the White House following the Medal of Honor ceremony. (White House Memorial Album)

96 Chapter 6: Medal of Honor

Daniel's family meets with President and Mrs. Johnson and other dignitaries in the White House following the Medal of Honor ceremony. (White House Memorial Album)

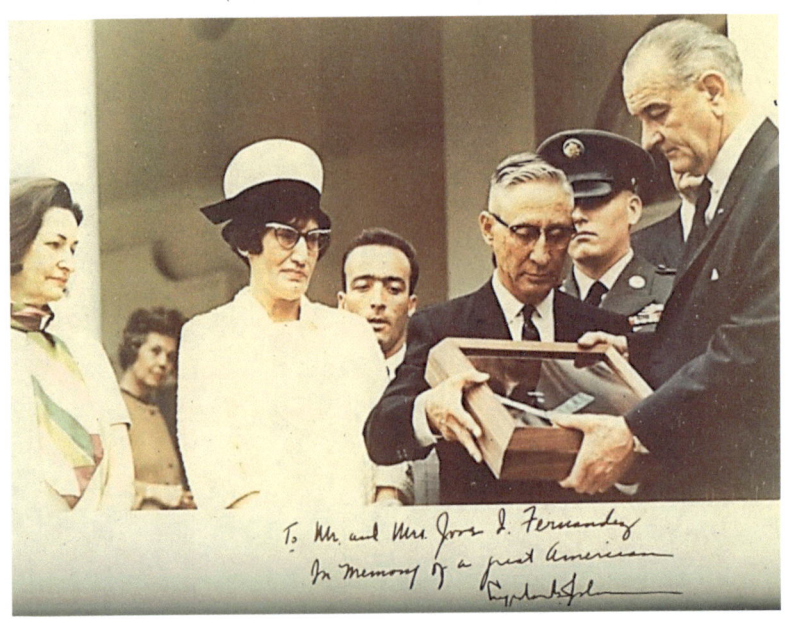

Signed by the President: "To Mr. and Mrs. José I. Fernández. In memory of a great American, Lyndon B. Johnson." (White House Memorial Album)

SGT Ray E. Sue, one of the men SP4 Daniel Fernández saved, meeting President Lyndon B. Johnson and Major Hugh Robinson, Presidential Aide, in the Diplomatic Reception Room at the White House, following the Medal of Honor ceremony. (White House Memorial Album)

James P. McKeown, another of the four men Daniel saved, meeting President Johnson at the White House following the Medal of Honor ceremony. (White House Memorial Album)

*Daniel's father, José, looks over Daniel's awards and medals.
(White House Memorial Album)*

*José and Laura Fernández looking at Daniel's awards and medals.
(White House Memorial Album)*

Daniel D. Fernández: The Man Behind the Medal of Honor 99

*Colonel Lynnwood M. Johnson Jr. speaks to the family and friends of Daniel Fernández during a luncheon at the Madison Hotel following the Medal of Honor presentation ceremony in Washington, D.C.
(White House Memorial Album)*

(Left to Right) James P. McKeown, James and Laura Fernández, Colonel Lynnwood M. Johnson, José, Peter, and Rita Fernández, and Father Francis Schuler. (White House Memorial Album)

100				Chapter 6: Medal of Honor

While in Washington, D.C., the Fernández family visited Arlington National Cemetery and President John F. Kennedy's Eternal Flame then attended Mass at the Basilica of the National Shrine of the Immaculate Conception Cathedral. (White House Memorial Album)

Chapter 7: Other Honors and Tributes

Many individuals, especially his brothers James and Peter and Peter's wife Lollie, have continued to keep Daniel's heroism and memory alive with presentations, interviews, and other public events. Moreover, there have been numerous tributes and honors paid by individuals and organizations across the country.

Other Military Honors

In addition to the Medal of Honor, Fernández received three Purple Heart awards, the Combat Infantryman Badge, the Marksmanship Badge, the National Defense Service Medal, the Vietnam Campaign Medal, the Vietnam Service Medal, the Army Presidential Unit Citation, the Valorous Unit Award, the Vietnam Gallantry Cross, the Air Medal, and the Army Good Conduct Medal, as featured below and in the portrait.

Medals Daniel received.

A portrait by Maximos was unveiled by the National Assessment Group at the rededication of Fernandez Hall at Kirtland Air Force Base. The portrait depicts how Daniel would have looked wearing his medals if he had survived the incident that took his life. (Fernández Collection)

Editorial
Valencia County News, February 1966

We Salute You, Daniel Fernández

A young man who hoped one day to become a heavy equipment operator died saving the lives of six [sic] fellow soldiers Friday in a combat zone area in Viet Nam. His name was Daniel Fernández, son of Mr. and Mrs. Jose I. Fernández of Los Lunas.

The soldier gave his life when he tossed his body upon an enemy grenade, an act for which he has been nominated for the Congressional Medal of Honor, the highest tribute this country can bestow upon a hero.

We will never know what impulse flashed through Daniel's mind in what must have been not more than a second from the time he saw the grenade and the blast itself. Whatever it was, the action was motivated by a trait found only in rare individuals.

We do know that his actions on the fatal Friday were typical of Daniel.

Once before under enemy fore, Daniel went to the aid of three "buddies." During this mission he was shot in the foot. He was later decorated with the Purple Heart.

Daniel did not have to go to Viet Nam the first time. He volunteered to go as a "shot-gun" rider on an Army helicopter. His job was to protect the vulnerable aircraft from enemy ground fire. His tool was a .50 caliber machine gun.

For his service is this capacity, Daniel was awarded the Air Medal.

When Daniel returned to Viet Nam January 18, 1966, he had been wounded once and decorated twice for previous duty in the Victnamese War.

From Jan. 18 to Feb. 18, the day on which he died, Daniel, with his outfit, was frequently under heavy fire. In one letter received by a friend in Los Lunas, shortly before his death, Daniel wrote that in an action a few days ago, his company lost 16 men.

Daniel planned to leave the Army after his four-year hitch. His father told us he was due to be discharged in November of this year.

But Daniel wasn't counting the days on Feb. 18, he was still a soldier doing the job the best he knew how.

We salute you, Daniel Fernández.

Statement by New Mexico Senator Joseph P. Montoya *Congressional Record of the 90th Congress*, Apr. 6, 1967

Senator Joseph P. Montoya

Thousands of boys like Danny Fernández are serving today, and I know so many of them. These young men are not on the picket lines. They are not advocating for peace at any price. They do not take LSD. They are not interested in destroying property, defying constitutional authority, or mocking their elders.

Rather they have a deep devotion to country, democracy, and loved ones. It is because they believe with all their hearts and minds that they go forth to answer to the call of their Nation. They go to serve, to fight, to be wounded, and some to die, far away from home and hearth. Far away from the sound of a mother's voice and a sweetheart's hand. Far away from the familiar and the comfortable. Far away from the blessings of our own land.

Boys like Danny Fernández love our land, our traditions, our dreams. They want a better life and a fuller realization of what they fervently believe they are fighting for. It is because their belief is translated into action that our Nation survives. We can suffer their loss and mourn them, but we cannot betray them.

Song: El Corrido de Daniel Fernández

Roberto Martínez was born in Mora County, New Mexico. After Roberto married, he and his wife, Ramona, moved to Denver in 1952. There he played in a musical duo with Jesse Ulibarri, his wife's uncle. They moved to Albuquerque in 1961 where Roberto found employment on Kirtland Air Force Base. In 1962, Roberto co-founded "Los Reyes de Albuquerque" with Ray Flores, Miguel Archibeque, and other friends. The band played traditional New Mexican music throughout the Albuquerque area. Their music thrived, but their popularity exploded onto the scene in 1966.

In February 1966, Roberto heard about the death of Daniel Fernández. He saw Daniel's parents on television newscasts and in the newspapers. He felt the loss of their son, a heroic young Hispanic man that gave his life to save the life of his fellow soldiers. Their parental sorrow and Daniel's deed moved Roberto so profoundly that it prompted him to write his first *corrido* as he was driving down the road one day.[5]

Los Reyes de Albuquerque performed Daniel's ballad live on KABQ radio. When Roberto returned home, the radio station called and told him he should record the *corrido* because the station's switchboard was flooded with calls after Los Reyes' performance.

Out of respect to the family, Roberto contacted the Fernández family, to ask for their permission to record the song. Initially, they declined, probably because their grief was still too painful. However, the family eventually agreed to let Los Reyes de Albuquerque record the song that was produced by famed New Mexican singer Alberto Sánchez whose stage name was Al Hurricane. Cynthia Shetter recalls

---

[5] A *corrido* is a traditional Mexican folk song genre that tells a story, frequently a sad one, in ballad form.

that she had to ask patrons at the Los Lunas Museum of Heritage and Arts to wear headphones whenever they listened to the *corrido* so that Daniel's brother, James Fernández, the Museum director, would not have to hear the sad music over and over during the 2008 "Men and Women of Valor Exhibition" featuring Daniel.

Roberto attributes his group's long-standing success to Daniel's *corrido*. They were soon asked to open for José Feliciano at the Palladium in Hollywood and Anita Bryant at the National VFW Convention in New York. They also had the opportunity to play and record the *corrido* with Mariachi Vargas de Tecalitlán. Los Reyes considered this a great honor because Mariachi Vargas of Tecalitlán, Jalisco, Mexico, was recognized as "The World's Greatest Mariachi" during the 1950s.

Miguel Archibeque and Roberto Martínez of Los Reyes de Albuquerque singing in studio. (Center for Southwest Research)

45 RPM record of El Corrido de Daniel Fernández, released in 1966. (Daniel D. Fernández Collection)

El Corrido de Daniel Fernández

En Español

Amigos vengo a cantarles
El corrido de un paisano;
Se llamo Daniel Fernández
Hijo Nuevo Mexicano.

Este soldado valiente,
Valiente de Nuestro estado;
Por el amor a su patria
La vida ha sacrificado.

Estribillo (Chorus):
Su vida fue terminada
Murió en batalla mortal;
Ahora se encuentra con Díos
En su reino celestial.

En el pueblo de Los Lunas
Fue el lugar donde nació;
Y en el Sur de Viet Nam
Fu el lugar donde murió.

Era grande de estatura
Y grande de corazón
Y a nuestra patria querida
Le sirvió con devoción.

Estribillo (Chorus):
Nuevo Mexico Querido
No des tu brazo a torcer;
Tienes soldados Valientes
Que cumplen su deber.

Decia este gran soldado
Cuando se vio mal herido;
Virgencita milagrosa
Nomas un favor te pido.

Dame un momento de vida
Para rezar un rozario;
Después madrecita mía
Contento me voy contigo.

Estribillo (Chorus):
Ya con esta me despido
Pasándome en estos valles;
Aquí termino el corrido
Del gran soldado Fernández.

English Version

Fellow Americans listen,
I would like your attention.
So, I can tell you a story,
Of love, beauty, and devotion.

A ballad of Dan Fernández
He is the title of this song.
He died in South Vietnam
While fighting a valiant cause.

I say hello to Los Lunas
He used to run, play, and cry.
Then on to South Vietnam
Where he was destined to die.

When he was mortally wounded,
A soldier had this to say,
Dear Lord, I am ready to go,
Just give me the time to pray.

Do say an act of contrition,
That is all I wish to do.
And then my heavenly Father,
I will gladly go with you.

Chorus:
The love in New Mexico,
You have the right to be proud.
One of your sons has distinguished,
Himself today in the crowd.

Scan the QR code below to hear *El Corrido de Daniel Fernández* by Los Reyes de Albuquerque:

Spanish   English

Spanish: https://www.youtube.com/watch?v=vkKk0IOiCBU
English: https://www.youtube.com/watch?v=sRtL8AhfBEo

Song: Out Here in Viet-Nam

Joe Green, a carpenter from Albuquerque, attended Daniel's rosary, funeral, and burial in Santa Fe although he had never met Daniel. When asked why he drove all those miles for someone he didn't know he replied, "I had to go because Fernández did the greatest thing a man could do – he gave his life for his fellow man."

Concerned that he had not contributed anything himself, Green, a deeply religious man, decided to dedicate a song he had composed in memory of Fernández. The song is written in the perspective of a soldier writing home to his girl and highlights the longing he feels for the war to end so he could come home. Danny or any soldier in Vietnam, regardless of their rank, could have written the letter. You feel the soldier drawing strength from the love of his girl and his prayers to God.

Green approached Daniel's parents who liked the song and gave their consent to dedicate the song to their son. Faced with the problem of trying to get the lyrics on a record and set to music, Green went to Linden Blaschke, the popular leader of the garage rock 'n' roll band Lindy and the LaVells known throughout the Southwest. Blaschke agreed to release Green's "Out Here in Viet Nam" ballad on his Lavette record label. He quickly set it to music and recorded it himself.

The tune gained regional popularity, along with Lindy and the Lavells other songs "You Ain't Tuff" and "Let It Be" that surface today on a number of anthologies of rare and classic garage 45 LP's.

110   Chapter 7: Other Honors and Tributes

Lindy Blaskey and the Lavells. (NM Music Commission)

45 RPM record of "Out Here in Viet Nam" sung by Lindy Blaskey. (Youtube.com)

Out Here in Viet Nam

Never was much at writing,
But thought I would give it a whirl.
I'd like forgiveness,
I hope you are still my girl.

We quarreled on the day I left,
I know I left you blue.
Funny how these many miles,
Have brought me closer to you.

Chorus:
Out here in Viet Nam
The fighting gets really tough.
Thanks, be the day when the Viet Cong,
Will say they've had enough.
They've had enough.

Many of our buddies are gone,
Gone and it's a shame.
One thing we can assure you,
They didn't die in vain.

Chorus:
Out here in Viet Nam
The fighting gets really tough.
Thanks, be the day when the Viet Cong,
Will say they've had enough.
They've had enough.

I met our company chaplain,
Just the other day.
I listened with respect,
And this is what he had to say.

God give us strength,
God give us strength.

Again, I ask for forgiveness,
I hope to see you soon.
Just you and I beneath
the stars, the sun, and the moon.

Chorus:
Out here in Viet Nam
The fighting gets really tough.
Thanks, be the day when the Viet Cong,
Will say they've had enough.
They've had enough.

So put in a good word for me,
To the Lord above.
If your prayer is granted,
I promise you true love.

Scan the QR code below to hear *Out Here in Vietnam* sung by Lindy Blaskey:

https://www.youtube.com/watch?v=UasF-Fz5gjI

Poem by Gerald Van Dyke

MSGT, U.S. Army, retired, Gerald M. Van Dyke was a former member of the 25th Infantry Division in Hawaii. When he heard of Daniel's death, it inspired him to write a poem. He sent it to the Superintendent of the National Cemetery in Santa Fe and sent it to José and Laura. Gerald owned Van Dyke Music in Cordell, Oklahoma. José asked Gerald to set it to music, which he did.

While Gerald's poetry, music, and manuscripts are in the archives of the University of Oklahoma and Oklahoma State University, a recording of the song has yet to be found.

Our Danny was a rare GI,
A soldier all the way;
Unselfish, cheerful, competent,
So loveable each day.

In ambush, bullets from Viet Cong,
Were flying all around.
And Danny, with four comrades,
Were suddenly pinned down.

Oh, Danny, our Danny,
A soldier all the way.
Oh, Danny, our Danny,
Who never looked for pay.

If we could be like Danny boy,
No one would count the cost.
But willingly we would give all,
That freedom be not lost.

Oh, Danny, our Danny, our Danny boy!
A hand grenade was seen to roll,
Among his group of four.
And Danny knew that it might mean,
Their lives might be no more.
So Danny leaped on that grenade,

Just as it did explode.

And gave his live to save his friends,
At the end of a Hero's Road.

Oh, Danny, our Danny,
A soldier all the way.
Oh, Danny, oh Danny,
Who never looked for pay.

If we could be like Danny Boy,
No one would count the cost.
But willingly we would give all,
That freedom be not lost!

Oh, Danny Boy, our Danny, Our Danny Boy!

New Mexico Legislative Memorials

In February 1967, shortly after the announcement that Daniel's Medal of Honor nomination had been approved, Belen Representative Boni Tabet introduced a resolution honoring Daniel, which passed the state legislature unanimously. His parents were present for the ceremony. Representative Tabet noted, "Danny was a boy who became a man in an instant."

In February 2016, at the request of the Valencia High School National Honor Society project students, a memorial recognizing the 50th anniversary of Daniel Fernández's death in Vietnam was introduced by all five Valencia County representatives (Alonzo Baldonado, Kelly Fajardo, W. Ken Martínez, Matthew McQueen, and Don Tripp) and was passed by the House (HM 93, Feb. 2016). The students requested a similar memorial of the NM Senate (SM 85), which was introduced and supported by all Valencia County senators (Ted Barela, Clemente Sanchez, and Michael Sanchez). Their efforts also resulted in a resolution in honor of all New Mexico Medal of Honor recipients.

Peter and Priscilla Fernández, VHS National Honor Society students, and their sponsor Laurie Kastelic at the 2016 New Mexico Legislative Session. (Laurie Kastelic Collection)

New Mexico State Capitol Wall of Honor

The Wall of Honor in the Rotunda of the New Mexico State Capitol stands in tribute to Daniel Fernández and 11 other New Mexicans who were awarded the nation's highest military honor - the Medal of Honor. State Senator Lidio G. Rainaldi of Gallup was instrumental in the passage of Senate Joint Memorial 20, which authorized the display. The Wall of Honor was dedicated on November 4, 2005, to recognize 12 men who are an integral part of New Mexico's legacy of military service that dates back 400 years. Since then, two more Medal of Honor recipients have been added.

Those honored were: PVT Joe P. Martínez (WWII), 1st LT Alexander Bonnyman, Jr. (WWII), CAPT Robert Scott (WWII), PVT Harold Moon, Jr. (WWII), PFC Alejandro R. Ruiz (WWII), PFC José F. Valdez (WWII), 2nd LT Raymond G. Murphy (Korea), CPL Hiroshi Miyamura (Korea), SSG Drew Dix (Vietnam), SP4 Daniel Fernández (Vietnam), LCPL Kenneth L. Worley (Vietnam), SSG Franklin D. Miller (Vietnam), W.O. Louis R. Rocco (Vietnam), and SFC Leroy Petry (Afghanistan). (Photo: Cynthia J. Shetter)

Buildings and Facilities

Daniel Fernandez School

A group of Los Lunas High School students filled 27 pages with signatures requesting that a new middle school that was currently under construction, now Century High School, be named after Daniel Fernández. Although there was some question at first because the School Board wanted to name the high school gymnasium for Daniel, the Board eventually approved the request, and the school was so named in April 1966.

The completion of a new high school in 1969 had school officials repurposing buildings. The Los Lunas Luna High School that Daniel attended became the Los Lunas Middle School and the new building pictured below became Daniel Fernandez Elementary and eventually Daniel Fernandez Intermediate for the 5th and 6th grades. It was located at 32 Sun Valley Rd. in Los Lunas.

Daniel Fernandez Intermediate School at 32 Sun Valley Rd. in Los Lunas. (Fernández Collection)

116 Chapter 7: Other Honors and Tributes

VFW Post 9676 Dedication

Veterans of Foreign Wars (VFW) Post 9676 in Los Lunas was dedicated in honor of Daniel D. Fernandez on May 8, 1966. The post is housed in the historic Atchison, Topeka, and Santa Fe Railway Depot, which is located on the south end of Daniel Fernandez Memorial Park.

*Julian Silva (at left) and Henry Wortman (at right) are pictured holding the Daniel D. Fernandez VFW Post 9676 flag in May 1966 in honor of Daniel who was nominated for the Congressional Medal of Honor at that time. (*Valencia County News, *1966)*

Daniel D. Fernandez Post 9676 makes its home in a historic rail depot. (Cynthia J. Shetter)

Several men from the VFW Post knew Daniel personally and served in Vietnam at the same time that he did. They have worked closely with staff from the Los Lunas schools to ensure that Daniel and veterans of all wars are remembered.

Over the years, Post 9676 has continually taught students about character, honor, and selflessness through Daniel's and other veterans' stories. They encourage the students to discover their voices through the Voice of Democracy and Patriot's Pen scholarship programs.

In 2016, to commemorate the 50th Anniversary of the awarding of the Medal of Honor to Daniel D. Fernández, the VFW Post sponsored "A Life of Honor" essay contest and encouraged students at Los Lunas and Valencia High Schools to participate.

The post headquarters includes a portrait of Daniel in uniform with his Medal of Honor and a bust sculpted in 2007 by students participating in the Daniel Fernandez Memorial Center Project under the direction of the Artist in Residence, Terry Duncan.

Painting and sculpture created by 5th and 6th grade students on display in the VFW hall. (Cynthia J. Shetter)

118   Chapter 7: Other Honors and Tributes

Members of Daniel D. Fernandez Post 9676 with students at Daniel Fernandez Intermediate School in 2005. (Laurie Kastelic Collection)

A Life of Honor Contest sponsored by VFW Post 9676 in 2016. (Fernández Collection)

Auditorium and Gym at the Los Lunas Hospital and Training School

Governor Jack Campbell dedicated the combination gymnasium-auditorium at the Los Lunas Hospital and Training School in honor of Daniel Fernández on November 30, 1966. "I am confident that the Daniel Fernandez Gymnasium will be a worthy memorial to the spirit of love and compassion which he so dramatically exemplified in his life and death," the governor said.

The Los Lunas Hospital and Training School closed its doors in 1997 but the building is still utilized for state offices on the campus.

Daniel Fernandez Gym at the former Los Lunas Hospital and Training School. (Photo: John Taylor)

Governor Jack Campbell shaking hands with José and Laura Fernández at the gymnasium's dedication. (Valencia County News)

Daniel Fernandez Memorial Park

The Los Lunas Junior Chamber of Commerce sponsored a park on fifteen acres of land leased from the New Mexico Prison Farm adjacent to State Route 314 to be dedicated to Daniel Fernández. It was partially funded by a concert in the Los Lunas High School gym featuring Harry James, one of the most popular bandleaders during the "Swing Band" era of the 1930s and 1940s. Ground was broken in March 1966, and Senator Clinton Anderson formally dedicated the park on June 29, 1967. Today the park is used for family picnics, community recreation, and events.

Headline in the newspaper when Fernández Park was dedicated in 1967. (Valencia County News)

Peter and Priscilla Fernández at Daniel Fernandez Memorial Park on the 50th anniversary of Daniel's death in 2016. (Albuquerque Journal)

Daniel Fernandez Memorial Park was redesigned in 2020. It features a new memorial plaque. (Cynthia J. Shetter)

Memorial plaque in Daniel Fernandez Memorial Park. (Cynthia J. Shetter)

Daniel Fernandez Recreation Center

Daniel Fernandez Recreation Center. (Photo: John Taylor)

Within Daniel Fernandez Memorial Park is the Daniel Fernandez Recreation Center, a youth and recreation center which includes a gymnasium, a fitness center, a modern weight room, and office space. The first phase began in 1990 with two bathrooms, outside handball courts, and an office. The second phase constructed the gymnasium in 1996. A second gymnasium was added in 2022 along with renovations to the memorial park.

Schofield Barracks

Schofield Barracks in Oahu, Hawaii, was established in 1908 and named for Lieutenant General John Schofield, the officer who first recognized the strategic importance of a forward base on Oahu. It became the home of the 25th Infantry Division and was Daniel's "home away from home" during his initial deployment and during his rest and recovery period between Vietnam deployments. Fernandez Hall in the Schofield Barracks headquarters building was dedicated in Daniel's memory on June 19, 1975.

In 2004, Major General Eric T. Olson, commanding officer of the 25th Infantry Division, and his staff created the Schofield Memorial Project while serving in Afghanistan. The $500,000 project was designed to honor soldiers who were killed in action while serving in Afghanistan and Iraq. It took on an even greater meaning by honoring fallen comrades, including Daniel Fernández, from all battles that the 25th had participated in over its storied history.

Daniel D. Fernández: The Man Behind the Medal of Honor 123

Camaren Ly, Daniel's great niece, at Fernandez Hall at the Schofield Barracks headquarters building in Hawaii. (Fernández Collection)

Peter Fernández, at Fernandez Hall, viewing a plaque in his brother's honor. (Fernández Collection)

Chapter 7: Other Honors and Tributes

The sculpture, "United by Sacrifice," designed by Lynn Weiler Liverton, was dedicated in 2006 to "the memory of Tropic Lightning soldiers who made the ultimate sacrifice in defense of their country. "Greater love has no one than this, to lay down one's life for their country."

"United in Sacrifice" memorial at Schofield Barracks in Hawaii. (Fernández Collection)

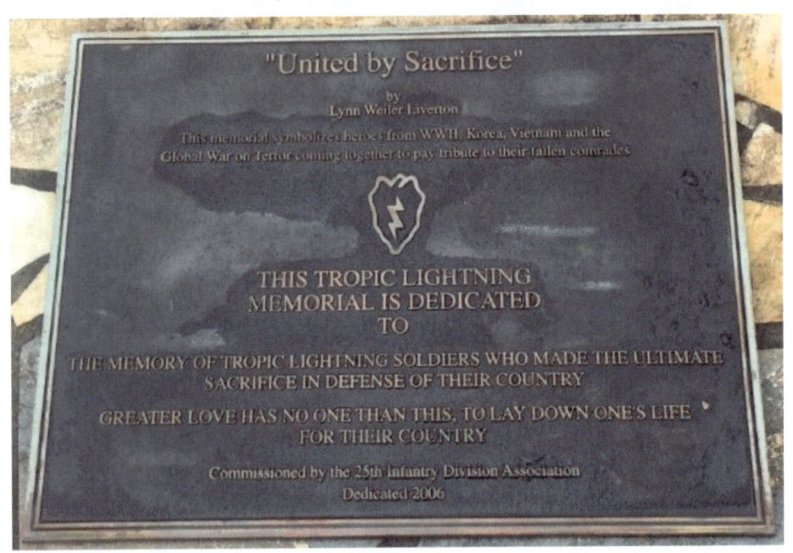

Dedication plaque for the "United by Sacrifice" memorial. (Fernández Collection)

In the 1970's, memorial bracelets were worn bearing the name of captured or lost soldiers. Now the bracelets are worn to keep loved ones, heroes, and victims alive in our hearts and minds. Here, Lollie Fernández, Daniel's sister-in-law, holds a metal memorial bracelet bearing Daniel's name that someone placed on the helmet, gun, and boot sculpture known as a Battle Cross at the "United by Sacrifice" memorial. (Fernández Collection)

Chapter 7: Other Honors and Tributes

Many memorial bracelets with Daniel's name are hung on the Battle Cross at the "United by Sacrifice" memorial. (Fernández Collection)

Camaren Ly, Daniel's grandniece, at the Schofield Barracks Memorial in Oahu, Hawaii. (Fernández Collection)

The 25th's 25th... In Combat

The 25th Infantry Division celebrated 25 years (October 1, 1941 – October 1, 1966) as an infantry division the year Daniel Fernández died. In 1966, the U.S. Army published a pictorial history book that spanned the division's actions in World War II, Korea, and the beginning of the Vietnam conflict. *The 25th's 25th ... In Combat* included stories of the valiant soldiers of World War II and Korea but utilizes Danny's death to exemplify the gallantry of the *Tropic Lightning* soldier.

> Those first weeks of battle were long and bloody, but not without the valor and heroism the 25th Infantry Division had carried through two previous wars.
>
> Infantrymen worked 24 hours a day, searching and clearing, digging foxholes, building bunkers-forcing the enemy to retreat farther and farther from his "sanctuary," depriving him of his supplies and his underground home.
>
> "Tropic Lightning" artillery and mortar rounds keep "Charlie" on the move. In many cases, the firepower from the Division's base camp has been so overwhelming that some of the enemy have completely given up and turned themselves over to the friendly forces. Tank and track crews advance through the jungles and swamps, clear roads and prove daily to "Charlie" their intent to stay until the mission is accomplished.
>
> During the short span of time, the Division has been in Vietnam, "Tropic Lightning" soldiers have carried the traditions of success in past wars to such places as the Boi Loi Woods, Ho Bo Woods, Bao Trai, Trung Lap and Trang Bang in the south, and Ban Me Thuot, Buon Brieng, Duc Co, and the Mangyang Pass in the north.
>
> There is a story of Danny Fernández a rare young man admired by his contemporaries – quiet, cheerful, competent, unselfish. While on a search and destroy mission with C Company, 1st Battalion, (Mechanized), 5th Infantry, Fernández and his buddies were in a battle position when a grenade came flying in. Fernández immediately jumped on it and covered it with his own body, shielding the others from the blast.
>
> Acts and deeds of heroism are recorded wherever the "Tropic of Lightning" operates in Vietnam.

Fernandez Hall at Kirtland Air Force Base

On October 17, 1984, the Fernández family was invited to Kirtland Air Force Base (KABF) for a dedication ceremony to honor the memory and heroism of Specialist Fourth Class Daniel Fernández, United States Army. Air Force Colonel John M. Reeves and 1st SGT James C. Williams addressed a sizable crowd to dedicate Building 20451. United States Representative Manuel Lujan Jr. was present and was instrumental in obtaining the federal funding for the building, which was designated to house the Command, Control, Communications Countermeasures Joint Forces (CCCCJF).

Air Force Colonel John M. Reeves addressing the audience at the 1984 dedication of Fernandez Hall on Kirtland Air Force Base in Albuquerque, New Mexico. (Fernández Collection)

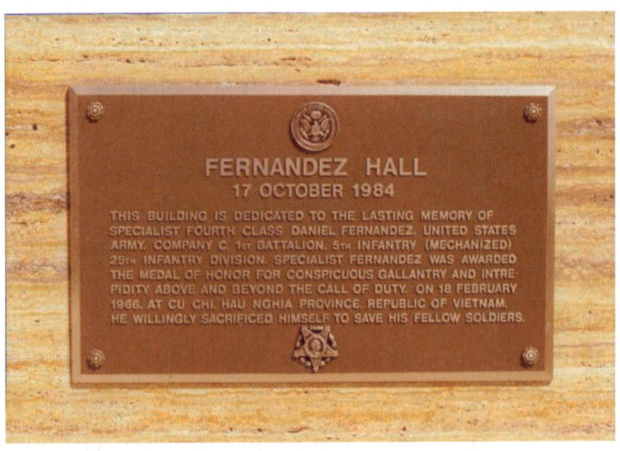

Dedication plaque for Fernandez Hall at Kirtland, AFB. (Fernández Collection)

Daniel D. Fernández: The Man Behind the Medal of Honor   129

Colonel John M. Reeves, Laura Fernández (Daniel's mother), and the director of CCCCJF cutting the ribbon at the dedication. (Fernández Collection)

James, Laura, Jason, Priscilla, John, and Peter Fernández in front of Fernandez Hall at Kirkland AFB in Albuquerque, New Mexico. (Fernández Collection)

Rededication of Fernandez Hall at KAFB

On October 6, 2017, members of the Peter Fernández family were invited to Kirtland Air Force Base for a rededication of Daniel Fernandez Hall as the centerpiece for the 20th anniversary celebration of the National Assessment Group, which tests technology for the Department of Defense and is housed in the hall.

Major General John P. Horner, director of special programs and the Defense Department Special Access Program Central Office, told family members, "It adds special meaning for this ceremony and for connecting us to your brother and hero SP4 Daniel D. Fernández."

Paul S. Greenhouse, the director of the National Assessment Group, was quoted as saying, "The memorial wall is our attempt to honor, in a more significant way, Daniel Fernández. We're trying to show Daniel is not just a memorial wall out there – Daniel Fernández is a person. He's a person that lived in New Mexico; he's a person with a family; he's a person that has personality and did wonderful things in his life. We're trying to portray that for everybody."

A soldier who saluted the fallen hero unveiled a portrait of Daniel Fernández, painted by Maximos. The painting depicts Daniel in his U.S. Army uniform with the medals that were bestowed upon his chest and the Congressional Medal of Honor around his neck.

One attendee, MSGT Tiffany Makris, grew up in Los Lunas and attended Daniel Fernandez Intermediate School. She was stationed at KAFB and worked within Fernandez Hall. Makris stated she felt a special connection to the war hero. "When I got assigned here two years ago and I walked past (Fernandez Hall) — I feel like this name follows me because I was deployed to Iraq and my living quarters were named Daniel Fernández," Makris said. She explained that they had several Alaskan tents to sleep in while in Iraq, each named after a Medal of Honor recipient. Makris was assigned to the Daniel Fernández tent.

"What we're trying to do here is step up our game, so to speak, and do a better job of representing his history but also personalizing it for members of this organization," Greenhouse said. "I think this is an extension of what they're trying to do in Los Lunas – they're saying, 'We're proud of our native son. We want people to know about him. We want other people to understand what a great community we have that produces people like Daniel Fernández and has families like the Fernández family.'

Paul S. Greenhouse, director of the National Assessment Group, addresses the Fernández family at the rededication of Fernandez Hall at Kirtland AFB, Albuquerque, New Mexico, in 2017. (Todd R. Berenger, NAG)

A soldier honoring U.S. Army Spec. Daniel D. Fernández at the rededication ceremony of Fernandez Hall on October 6, 2017. (Todd R. Berenger, NAG)

132　　　　　　　　　　　　Chapter 7: Other Honors and Tributes

Major General John P. Horner (2nd from left) and other military dignitaries present Peter and Lollie Fernández with a copy of the portrait of Daniel. (Dennis Carlson)

Lollie and Peter Fernández in front of Fernandez Hall at Kirkland AFB, Albuquerque, New Mexico. (Fernández Collection)

Memorials and Monuments
Vietnam Wall

On November 13, 1982, a "V" shaped Vietnam Veterans Memorial wall, designed by architect Maya Lin, was dedicated in Constitution Gardens in Washington, D.C. Seventy separate panels make up each section of the 246-foot-9-inch-long memorial that is composed of black granite. When dedicated, 57,939 names of service members who were classified as dead, missing, or prisoners of war were engraved in chronological order by the date of their deaths. They begin and end at the center of the memorial where the two sections meet. Having the names begin and end at the center is meant to form a circle – signifying a completion to the war. Additional names have been added throughout the years, making the name count over 58,000, joining Daniel Fernández's name on panel 5E, line 46 on the Wall.

On Veterans Day 1996, the Vietnam Veterans Memorial Fund unveiled a mobile replica of the Vietnam Veterans Memorial in Washington, D.C. Bringing "The Wall" home to communities throughout the country is a reminder of these service members' sacrifice and for their family and friends to find peace and comfort in familiar surroundings.

Since its inception in 1996, more than one million people have visited the Memorial Funds Traveling Wall exhibitions. Traveling Walls have made stops in nearly 200 U.S. locations.

In February 2003, one of the traveling replicas of the Vietnam Veterans Memorial Wall made its permanent home in Truth or Consequences, New Mexico. The Truth or Consequences wall traversed the country for approximately three years before being retired in December 2002. The half-scale model was purchased by the State of New Mexico with the assistance of local businessmen and installed in Veteran's Memorial Park in Truth or Consequences.

One of the "portable" walls was on display in Daniel Fernandez Memorial Park in Los Lunas on May 12, 2004. Daniel's brother, Peter, along with Elijah Griego, Gloria Chávez-Simpson, Brenda Howard, Lydia Piro, Fermin Monavica, and Claire Maestas read the names of the men who gave their lives during this conflict. The bell ringers included Katarina Griego, Fermin Castillo, Marlinda Valdez, Grady Rigdon, Priscilla Fernández, and Mike Maestas.

Chapter 7: Other Honors and Tributes

These memorials continue to be a symbol of enduring legacy, healing, and education.

Emma and Garrett Fernández, Daniel's grandniece and grandnephew, visiting the Vietnam War Memorial in Washington, D.C., 2018. (Fernández Collection)

Daniel D. Fernández: The Man Behind the Medal of Honor 135

Daniel's grandnephew, Zachary Fernández, making a rubbing of Daniel's name. (Fernández Collection)

Rubbing of Daniel's name from the memorial wall. (Fernández Collection)

Half-scale model of Vietnam Memorial in Truth or Consequences, New Mexico. (Photo: Richard Melzer)

Valley Forge, Pennsylvania, Medal of Honor Grove

New Mexico obelisk in the center of trees planted in 1974 to honor 12 New Mexico Medal of Honor recipients, including Daniel Fernández. (Freedoms Foundation)

The Medal of Honor Grove at Valley Forge, Pennsylvania, was conceived by Kenneth Wells, General Dwight D. Eisenhower, E.F. Hutton, General Omar Bradley, and others at Freedoms Foundation. The site was dedicated on the Freedoms Foundation's campus in 1964.

The Grove, 42 acres of natural woodland, is the oldest living memorial honoring the more than 3,500 Medal of Honor recipients. An area of land is set aside for the recipients in each of the 50 states, as well as Puerto Rico and the District of Columbia.

Each state has a seven-foot obelisk centered on a 25-square-foot brick plaza. The recipients accredited by the Congressional Medal of Honor Society to that state are identified by name, rank, and service branch on the obelisk. Recipients are additionally honored with a ground marker engraved with their names, branch of service, and the date and location of their acts of valor.

On September 14, 1974, Daniel Fernández and the other New Mexico Medal of Honor recipients were memorialized in the Grove. They are Alexander Bonnyman, Delbert Jennings, Joe P. Martínez, Franklin Miller, Hiroshi Miyamura, Harold Moon Jr., Francis Oliver, Eben Stanley, José Valdez, Kenneth Walker, and Kenneth Worley.

New Mexico First Lady Alice King said at the event:

> It is fitting and proper that we should forever memorialize these men by setting aside this small but beautiful area, where trees can grow, where leaves can fall, where birds can sing and nest, and where it can be forever quiet and peaceful. There is a certain sanctity about this place, where the spirits of these brave men and the memory of their deeds will unite with the spirits of their predecessors in the Continental Army and will pervade the consciousness of all who visit here. The acts for which these men were honored were born in turmoil and agony, and it seems most fitting that they can be commemorated in the quiet and peacefulness of this place.
>
> As we dedicate this shrine today for the people of New Mexico and the United States, I hope we will always remember that the few men whom we honor today are truly representative of all men through the ages who have willingly sacrificed themselves for the benefit of their fellow man. I would like to close these brief remarks by offering each of our Medal of Honor recipients, both present and absent, and to all Medal of Honor recipients who are and will be honored in these groves, a fervent prayer that the sacrifices which these men so selflessly performed will never go unnoticed or unheeded, and that the true message of their selfless sacrifice will be forever enshrined in the hearts of all people forever.

138   Chapter 7: Other Honors and Tributes

Marker at the Medal of Honor Grove at Valley Forge, Pennsylvania. (Hazel Willis Collection)

Men from Daniel Fernández's unit visiting the memorial for "Danny," as they knew him, at Valley Forge during 25th Infantry Division Reunion in 2006. (Hazel Willis Collection)

25th Infantry Division Reunion at the Medal of Honor Grove at Valley Forge, Pennsylvania, in 2006. (Hazel Willis Collection)

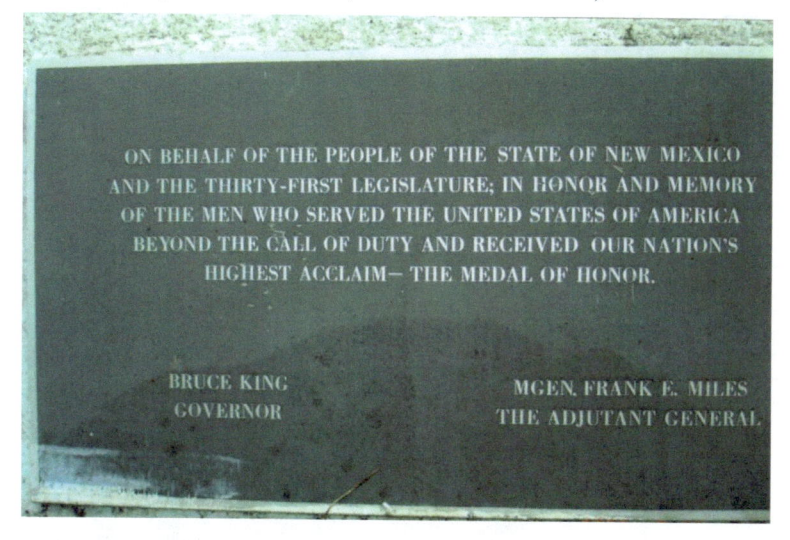

Plaque dedicating the memorial by Governor Bruce King and the Adjutant General Major General Frank E. Miles on behalf of the people of New Mexico, honoring the New Mexico Medal of Honor recipients at the Medal of Honor Grove in Valley Forge, Pennsylvania. (Hazel Willis Collection)

Medal of Honor Museum within Patriot's Point Naval & Maritime Museum Charleston, South Carolina

The Congressional Medal of Honor Society Museum is home to the Medal of Honor, the recipients who wear it, and the values inherent in the Medal - courage, commitment, integrity, sacrifice, citizenship, and patriotism.

Designed to bring the recipients' stories to life, the museum is located on board the USS *Yorktown* (CV-10) at Patriot's Point Naval & Maritime Museum in Mount Pleasant, South Carolina. Interactive exhibits focus on recipients' stories from various wars. A searchable database allows visitors to explore all Medal of Honor citations. Permanent exhibits of artifacts tangibly connect the past and present.

The last gallery focuses on how a recipient's moment of valor can have a lasting impact on the United States' story. It is a moment for visitors to reflect and gather further inspiration on how their own lives can make a difference, on how embracing the Medal of Honor values can influence those around them and the world we live in.

Commissioned in 1943, the USS Yorktown *(CV-10) was the tenth aircraft carrier to serve in the U.S. Navy. Nicknamed the "Fighting Lady", the Yorktown served the nation for almost three decades before being decommissioned in 1970 and becoming the first ship of the Patriots Point fleet when the museum opened in 1976. (Hazel Willis Collection)*

Daniel D. Fernández: The Man Behind the Medal of Honor   141

Kiosk at the Congressional Medal of Honor Museum on board the USS Yorktown *where visitors can look up Medal of Honor Recipients. (Hazel Willis Collection)*

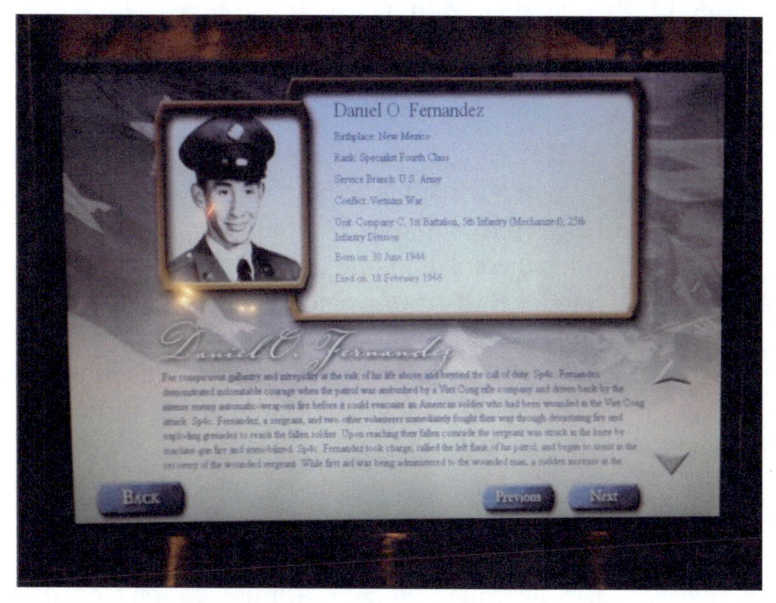

The Medal of Honor Museum's Interactive exhibit allows visitors to search for recipients including Daniel. (Hazel Willis Collection)

Blue Spruce planted at the Veterans' Administration Hospital in Albuquerque

On October 25, 1976, a 16-foot "Bicentennial tree," a blue spruce, was planted at the Veterans Hospital in Albuquerque to honor Daniel Fernández and eight other Medal of Honor winners. The other men honored were:

- Retired Col. Robert S. Scott of Santa Fe, New Mexico, for single-handedly holding off an enemy attack in New Georgia, Solomon Islands during World War II.
- Maj. James Fleming, helicopter pilot, from Sedalia, Missouri, for rescuing a patrol in the face of enemy fire during the War in Vietnam.
- Marine Lt. Raymond G. Murphy for leading an evacuation up a hill, although wounded, during the Korean War.
- Marine Maj. Jay R. Vargas from Winslow, Arizona, led a charge on an enemy position in Vietnam in 1968.
- Army Cpl. Hiroshi H. Miyamura of Gallup, New Mexico, killed 10 men in hand-to-hand combat in one fight during the Korean War.
- Army Sgt. Alejandro R. Ruiz of Lovington, New Mexico, killed 12 of the enemy in a fight for a pillbox on Okinawa, in 1945.
- Marine 1st Lt. Alexander Bonnyman Jr., died in the Gilbert Islands in 1943 attacking a gun emplacement after being responsible for 250 enemy deaths.
- Army Lt. Robert McDonald won a medal for action at Wolf Mountain, Montana, in 1877.

Air Force Maj. James Fleming, Army Col. Robert S. Scott, and Marine Lt. Raymond G. Murphy share in the tree-planting. (Albuquerque Tribune).

Along with the tree planting, a plaque was placed under the blue spruce tree at the Albuquerque Veteran's Administration Hospital. Unfortunately, the tree was removed in 2012 during construction and the plaque has since disappeared.

Display at the Veterans Memorial in Albuquerque

An exhibit featuring Daniel Fernández is on display in the building at the Veterans Memorial in Albuquerque. It features one of Daniel's three Purple Heart awards, a copy of Daniel's Medal of Honor citation, and a unique portrait of Daniel, painted by local artist Rafael E. Gonzales.

Gonzales served in the U.S. Army 167th Signal Corp. During his combat tour in Vietnam he sustained direct trauma to his eyes causing severe visual impairment. Through art, Gonzales learned to channel the emotional effects of his visual condition and combat experience.

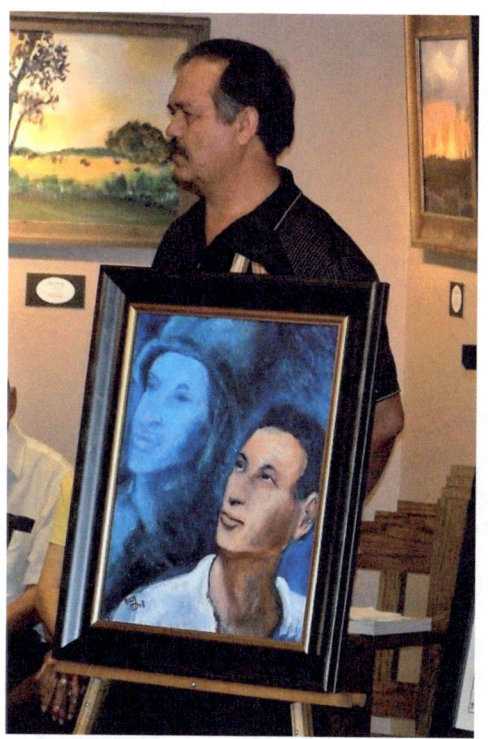

Daniel Fernández's Purple Heart. (Laurie Kastelic Collection)

Rafael E. Gonzales presenting the portrait of Daniel Fernández he painted, at the opening of the Men and Women of Valor Exhibition at the Los Lunas Museum of Heritage & Arts in July 2008. (Laurie Kastelic Collection)

Exhibit at National Hispanic Cultural Center

An exhibit entitled "New Mexico Hometown Heroes: Hispanic Medal of Honor Winners," was opened on Thursday, November 8, 2018, at the National Hispanic Cultural Center in Albuquerque. This exhibit, which ran through January 2019, honored Fernández along with Joe P. Martínez (World War II), Leroy Petry (Global War on Terrorism), Louis Rocco (Vietnam), Alejandro Ruiz (World War II), and José Valdez (World War II).

SP4 Daniel D. Fernández

PVT Joe P. Martínez

SSGT Leroy Petry

SGT1C Louis Rocco

PFC Alejandro Ruiz

PFC José Valdez

Chapter 8: Legacy of a Hero

On February 18, 1966, Daniel Fernández's name was added to a list of some 400 New Mexicans who died in the Vietnam Conflict. His death riveted a nation and brought the reality of the growing conflict home. It was alternately heartbreaking and pride-inducing. Today, most historians would agree that Daniel remains arguably the most famous New Mexico soldier to give his life during the Vietnam Conflict.

The Vietnam Conflict was a pivotal chapter in the history of the Cold War. The overall mission of the United States military forces was to stop communism from spreading in Southeast Asia. On November 14, 1963, President John F. Kennedy gave his last White House Press Conference on the situation in Southeast Asia, highlighting areas held by communist rebels just one week before he was assassinated on November 22, 1963. Nevertheless, the 20-year-long war (1955-1975) occurred at the same time as civil unrest and its political implications that were taking place in the United States.

Today, Vietnam veterans are recognized and treated with more respect than when they initially returned from the war. Their service is appreciated but some historians, academics, and veterans believe there is a disconnect with the youth of today in understanding not just their contribution but veterans' contributions overall.

Dr. Richard Melzer, former University of New Mexico professor, felt that Daniel's story was an important element to include in the telling of New Mexico's role in the Vietnam Conflict in his 2011 book, *A History of New Mexico Since Statehood*, that is used as a high school level textbook throughout New Mexico. In the book, Daniel's story accompanies the 1970 anti-war protest in Albuquerque - another decisive event in New Mexico history.

Members of the Fernández family have been visiting schools for over four decades to help teach the students not only about Daniel but to highlight the legacy of veterans in general. Daniel's brother, Peter, has been visiting Los Lunas schools for several years talking with the students about "What is a hero?" He stated in a 2019 *Santa Fe New Mexican* article, "They know of a park, school, and street named after him. What he did, they did not know." When asked, "What is hero?" the boys respond, "Michael Jordan" and the girls respond, "Beyoncé."

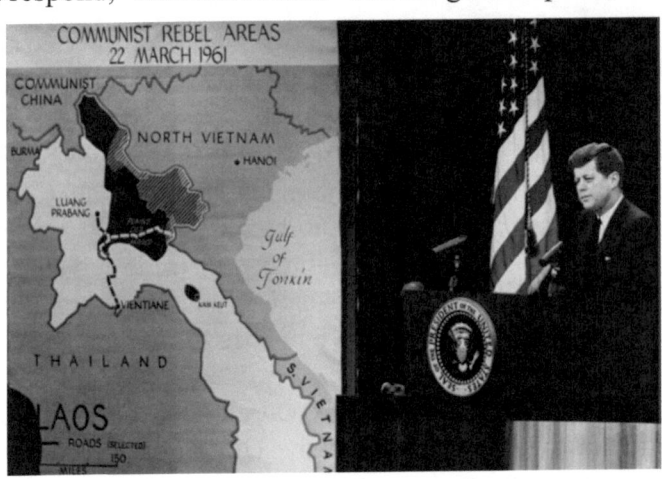

President John F. Kennedy brief media on the situation in Southeast Asia one week before his death in 1963. (JFK Presidential Library and Museum)

Daniel Fernández Intermediate (DFI) School librarian Laurie Kastelic came to a similar conclusion. Realizing that her students didn't even know why their school was named after Daniel Fernández, she embarked on a program to change that.

In Spring 2005, 12 sixth-grade students at DFI in Los Lunas researched Daniel's life and service with the help of the Fernández family and numerous veterans who knew Daniel, both locally and from his infantry division in Vietnam. The students wrote a letter to the *Valencia County News-Bulletin* editor asking people that knew Daniel to contact them. They signed the letter Vanessa Aragon, Marissa Carrasco, Ashley Curley, Heather Drummond, Matthew Fleischer, Nicole Gurulé, Veronica Mendoza, Sarah Miller, Rosie Simmons-Hogan, and Mrs. Laurie Kastelic, Coordinator, Daniel Fernandez Intermediate School.

Armed with the photos, stories, and artifacts they collected, the students created a permanent educational display that was installed in the school's library and subsequently dedicated in June 2005 as the Daniel Fernández Memorial Center display. Two years later, in Spring 2007, a group of DFI fifth graders researched and developed a five-page website that included Daniel's story and information on the Medal of Honor. The site was housed on the DFI school website for the next three years.

In Spring 2009, a class of four fifth and sixth grade students researched, wrote, and illustrated a children's book, *Man of Honor: The Story of Daniel Fernández*, published by AuthorHouse. In addition to Daniel's story, the book included a history of the three student projects to that point. A reception was held in late May 2009, at which the students presented copies of the new book to the Fernández family and local veterans.

VFW Post 9676 Commander Larry Artiaga (on right) showing DFI students a map of Vietnam. (Laurie Kastelic Collection)

Peter Fernández telling DFI students about Daniel. (Laurie Kastelic Collection)

148   Chapter 8: Legacy of a Hero

DFI students and Daniel's brother Peter viewing the NBC News footage of the day Daniel died. (Laurie Kastelic Collection)

Educational display the DFI students created. The plaque on the right reads: This display is dedicated to the memory of Daniel Fernández, a young man who knew the true meaning of CHARACTER. The Daniel Fernández Memorial Project May 2005. (Laurie Kastelic Collection)

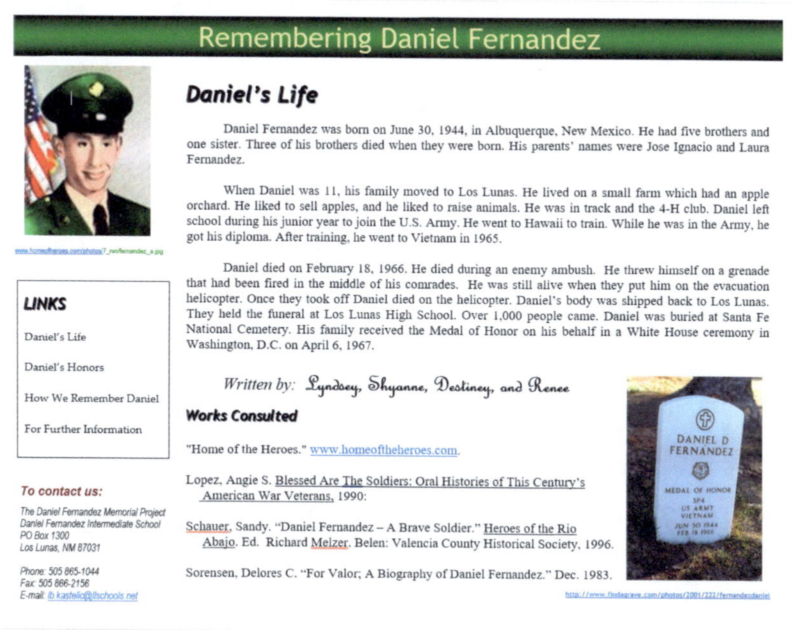

Website on Daniel's life created by DFI students in 2007. (Laurie Kastelic Collection)

Website about Daniel in Vietnam created by DFI students in 2007. (Laurie Kastelic Collection)

150   Chapter 8: Legacy of a Hero

Mrs. Bouts, Katie Kasky, Baylee Glass, Enrique Villarreal, and Joseph Oroña, the DFI students who wrote and illustrated Man of Honor, *with Lollie Fernández, Daniel's sister-n-law. (Laurie Kastelic Collection)*

Peter, Lollie, and James Fernández with the DFI authors of Man of Honor: The Story of Daniel Fernández. *(Valencia County News-Bulletin)*

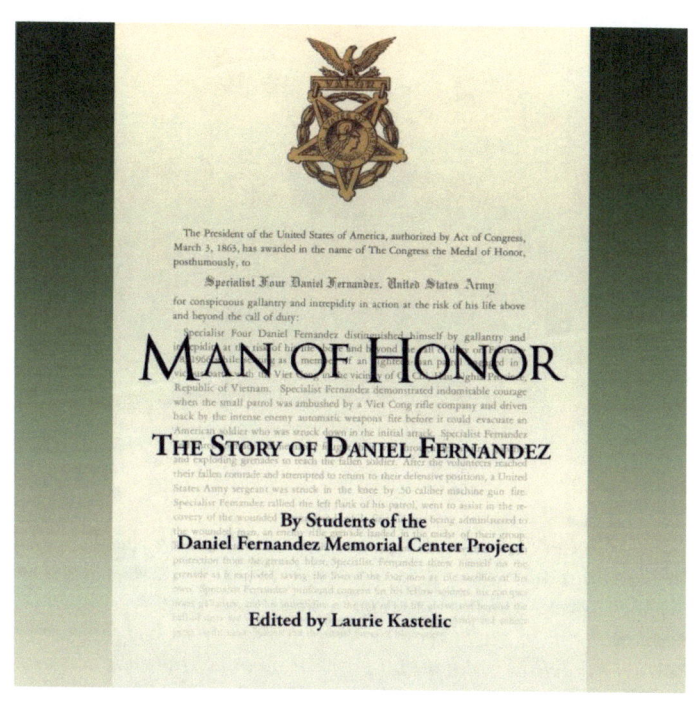

Daniel D. Fernández: The Man Behind the Medal of Honor

The book created by DFI students and edited by Project Coordinator Laurie Kastelic is available on Amazon.com.

Text from Man of Honor: *"On Daniel's first tour of Vietnam, he was a soldier whose job was to fight the enemy from a helicopter. Daniel was hurt in combat and was sent to Hawaii to heal."*

Chapter 8: Legacy of a Hero

Text: "On February 18, 1966, on his second tour in Vietnam, Daniel was on a mission to help a wounded soldier .... As Daniel and his patrol were trying to rescue the soldier, the enemy threw a grenade at them. In the hurry to get away, Daniel kicked the grenade toward his fellow soldiers by mistake. Daniel saw that his patrol couldn't get away in time, so he leaped on the grenade and covered it with his body."

Text from Man of Honor: *"Daniel was just an ordinary farm boy who loved growing up in the country. Never in a million years would he ever have guessed that he would become a hero in his own hometown and even be remembered around the world!"*

Daniel D. Fernández: The Man Behind the Medal of Honor   153

During the 2007-2008 school year, the students from Daniel Fernandez Intermediate School published a folio of 16 charcoal portraits of Daniel for his brother, James Fernández. Here are three representative drawings:

*Drawings of Daniel Fernández by students at DFI.
(Laurie Kastelic Collection)*

Reunions

Over the years, members of Daniel Fernández's company and platoon have held several reunions where members of the company can reconnect with one another, and honor fallen members. On July 12, 2008, members of 1st Battalion, 5th Infantry Regiment 25th Infantry Division traveled to Los Lunas to hold a reunion after corresponding with Laurie Kastelic and the students at Daniel Fernandez Intermediate School. Norman "Butch" Pettis was instrumental in arranging the platoon's reunion that July in Los Lunas after he saw what Kastelic and her students were trying to accomplish.

Working in coordination with Butch, the Fernández family, the Los Lunas Schools, VFW Post 9676, the ROTC, the Village of Los Lunas, the Parks and Recreation Department, and the Museum of Heritage & Arts, Laurie Kastelic arranged a series of events to welcome the reunion attendees. The events included a visit to DFI, a ceremony at the Daniel Fernandez Memorial Park, and the opening of the "Men and Women of Valor" exhibit at the Los Lunas Museum of Heritage & Arts featuring photos from Daniel Fernández's photo albums.

Over two dozen veterans and their families from the 1st Battalion, 5th Infantry Regiment, 25th Infantry Division attended the reunion. They began their day at Daniel Fernandez Intermediate School where the students and staff showed them the Daniel Fernandez Memorial Center and the projects they had been working on. The students also showed them the three-minute NBC News video highlighting the platoon moments after the attack that took Daniel's life. Tears rolled from the men's eyes as they watched. The scenes evoked emotions for those who had lost so much and could never forget.

The group then traveled to Daniel Fernandez Memorial Park where the Village of Los Lunas and the VFW Post 9676 hosted the veterans. The Los Lunas High School JROTC color guard was on hand as "Taps" was played, and the military guard gave a 21-gun salute.

Top Martínez, a childhood friend and a Vietnam veteran, told the crowd about growing up with Daniel and that he named his son after Daniel. He was now a grown man and a Marine.

Rick Stoeckley of Fort Wayne, Indiana, told those gathered that he met Daniel in 1965 while they were both stationed at Schofield Barracks in Hawaii. As a Private First Class (PFC) living in Hawaii it could be very expensive. He recalled that he bought a watch one payday and two weeks later he was flat broke. "I tried to sell it," he said as tears welled up in his eyes. "I wanted $20 for it, and Danny gave me $25. He said I should try and manage my money better."

Stoeckley said he goes to his small Catholic church in Fort Wayne, lights a candle, and says a prayer for Daniel. While kneeling in front of an altar, he thinks about his wife, his two sons, and the life he shouldn't have had. "I think about him every day."

He wonders if he is worthy of what Fernández did for him and the others assigned to the 25th Infantry Division. Stoeckley stated he has never forgotten the day when his friend died. "We had just come back in off our ambush and were just getting ready to break down our weapons," he said. "They were heavily pinned down, and one guy came back and told us what happened. We were the reactionary force, so we got our gear and went out. We all knew what happened before we got there."

"You wouldn't expect anything less," said Stoeckley of what Fernández did that day. "He was just that kind of guy. He was reserved. He wasn't going out and getting drunk like the rest of us. You could tell he was special. I'm glad I got to know him, and I'm glad he got the Congressional Medal of Honor. He deserved it. I will never forget him and the others who died for me over there. They run in my veins."

Dana Riley attended the reunion as well. He never talked about what he went through in Vietnam and the sacrifice Daniel Fernández made for him. It was too hard. Riley suffered from post-traumatic stress disorder (PTSD), and most of the country was reluctant to recognize Vietnam Conflict veterans, to say the least.

Riley finally confided in his daughter, and she helped him reunite with several men assigned to the 25th Infantry Division. That Saturday in Daniel Fernandez Memorial Park, Riley took to the stage and told the crowd of about 200 about Fernández.

"We carry our fallen brothers everywhere we go in our hearts." Riley said. "About 85 percent of the time, they are in our minds. It's just hard to talk about sometimes." Getting in contact with his fellow soldiers from the 1st Battalion, 5th Infantry 25th Infantry Division Class of 1966 helped Riley be able to cope with the harrowing events he witnessed in Vietnam.

State of New Mexico Cabinet Secretary of Veterans Affairs John García, a Vietnam veteran himself, was present that day. He chose to read a poem written by Archibald MacLeish:

The Young Dead Soldiers Do Not Speak

Nevertheless, they are heard in the still house,
Who has not heard them?
They have a silence that speaks for them at night,
And when the clock counts.
They say, We were young. We have died. Remember us.
They say, We have done what we could but until it is finished,
It is not done.
They say, We have given our lives but until it is finished,
No one can know what our lives gave.
They say, Our deaths are not ours: they are yours,
They will mean what you make them.
They say, Whether our lives and our deaths were for peace
and a new hope or for nothing we cannot say,
It is you who must say this.
They say, Whether we leave our deaths,
Give them their meaning,
Give them an end to the war and true peace,
Give them victory that ends the war and a peace afterwards,
Give them their meaning.
We were young, they say. We have died. Remember us.

García went on to say, "May we always be a grateful nation respecting the heroic sacrifices of our American heroes like Danny Fernández who served in harm's way to protect our country and our way of life. And may God give his family and his friends the comfort that is at need. May God always bless America."

The men of the 25th Infantry also honored Dan R. Shearin and Armando Tesillo of Bravo Company, the first members to be killed in action on January 31, 1966. Shearin's brother Ralph and sisters Hazel Willis and Holly Pruitt were on hand to accept a plaque in his honor.

After the ceremonies in the park, the group gathered at the Los Lunas Museum of Heritage & Arts for the opening of the "Men and Women of Honor Exhibit," curated by James Fernández, Daniel's brother. The exhibit featured photographs from four of Daniel Fernández's photo albums with photographs from Hawaii and his first tour of Vietnam.

Hazel Willis said, "It was shocking to come from Virginia, walk into the museum, and see pictures of our brother that we had never seen."

Daniel D. Fernández: The Man Behind the Medal of Honor 157

Chet Pino and James Garley from VFW Post 9676 presenting the post's National Citizenship Education Teacher Post Recognition Award to Laurie Kastelic for promoting citizenship in the classroom. (Laurie Kastelic Collection)

Rick Stoeckley, a member of 1st Battalion, 5th Infantry Regiment 25th Infantry Division, served in Vietnam with Daniel Fernández. (Valencia County News-Bulletin)

Members of the 1st Battalion, 5th Infantry, 25th Infantry Division with Laurie Kastelic at Daniel Fernandez Intermediate School. (Laurie Kastelic Collection)

Norman "Butch" Petit next to Laurie Kastelic, who was instrumental in organizing the July 12, 2008, reunion, seen here with other members of Daniel's platoon. (Laurie Kastelic Collection)

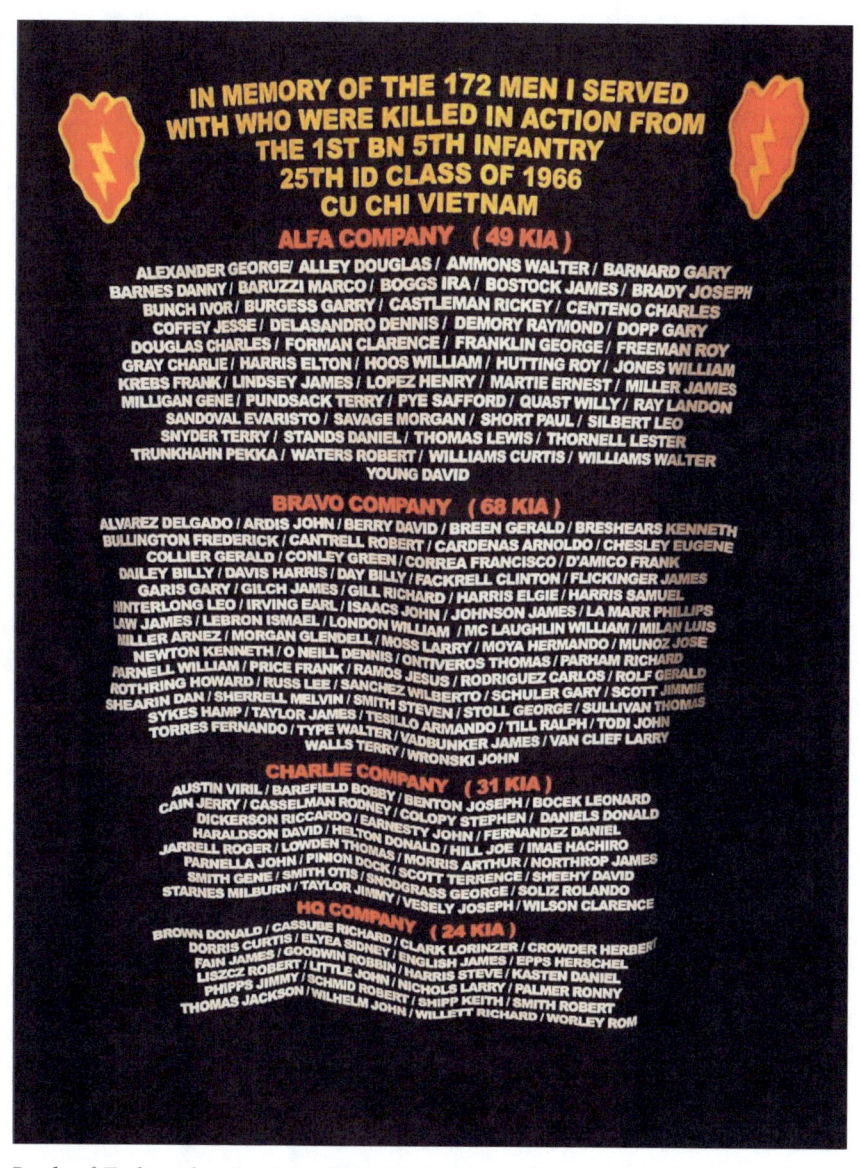

IN MEMORY OF THE 172 MEN I SERVED WITH WHO WERE KILLED IN ACTION FROM THE 1ST BN 5TH INFANTRY 25TH ID CLASS OF 1966 CU CHI VIETNAM

ALFA COMPANY (49 KIA)
ALEXANDER GEORGE / ALLEY DOUGLAS / AMMONS WALTER / BARNARD GARY
BARNES DANNY / BARUZZI MARCO / BOGGS IRA / BOSTOCK JAMES / BRADY JOSEPH
BUNCH IVOR / BURGESS GARRY / CASTLEMAN RICKEY / CENTENO CHARLES
COFFEY JESSE / DELASANDRO DENNIS / DEMORY RAYMOND / DOPP GARY
DOUGLAS CHARLES / FORMAN CLARENCE / FRANKLIN GEORGE / FREEMAN ROY
GRAY CHARLIE / HARRIS ELTON / HOOS WILLIAM / HUTTING ROY / JONES WILLIAM
KREBS FRANK / LINDSEY JAMES / LOPEZ HENRY / MARTIE ERNEST / MILLER JAMES
MILLIGAN GENE / PUNDSACK TERRY / PYE SAFFORD / QUAST WILLY / RAY LANDON
SANDOVAL EVARISTO / SAVAGE MORGAN / SHORT PAUL / SILBERT LEO
SNYDER TERRY / STANDS DANIEL / THOMAS LEWIS / THORNELL LESTER
TRUNKHAHN PEKKA / WATERS ROBERT / WILLIAMS CURTIS / WILLIAMS WALTER
YOUNG DAVID

BRAVO COMPANY (68 KIA)
ALVAREZ DELGADO / ARDIS JOHN / BERRY DAVID / BREEN GERALD / BRESHEARS KENNETH
BULLINGTON FREDERICK / CANTRELL ROBERT / CARDENAS ARNOLDO / CHESLEY EUGENE
COLLIER GERALD / CONLEY GREEN / CORREA FRANCISCO / D'AMICO FRANK
DAILEY BILLY / DAVIS HARRIS / DAY BILLY / FACKRELL CLINTON / FLICKINGER JAMES
GARIS GARY / GILCH JAMES / GILL RICHARD / HARRIS ELGIE / HARRIS SAMUEL
HINTERLONG LEO / IRVING EARL / ISAACS JOHN / JOHNSON JAMES / LA MARR PHILLIPS
LAW JAMES / LEBRON ISMAEL / LONDON WILLIAM / MC LAUGHLIN WILLIAM / MILAN LUIS
MILLER ARNEZ / MORGAN GLENDELL / MOSS LARRY / MOYA HERMANDO / MUNOZ JOSE
NEWTON KENNETH / O NEILL DENNIS / ONTIVEROS THOMAS / PARHAM RICHARD
PARNELL WILLIAM / PRICE FRANK / RAMOS JESUS / RODRIGUEZ CARLOS / ROLF GERALD
ROTHRING HOWARD / RUSS LEE / SANCHEZ WILBERTO / SCHULER GARY / SCOTT JIMMIE
SHEARIN DAN / SHERRELL MELVIN / SMITH STEVEN / STOLL GEORGE / SULLIVAN THOMAS
SYKES HAMP / TAYLOR JAMES / TESILLO ARMANDO / TILL RALPH / TODI JOHN
TORRES FERNANDO / TYPE WALTER / VADBUNKER JAMES / VAN CLIEF LARRY
WALLS TERRY / WRONSKI JOHN

CHARLIE COMPANY (31 KIA)
AUSTIN VIRIL / BAREFIELD BOBBY / BENTON JOSEPH / BOCEK LEONARD
CAIN JERRY / CASSELMAN RODNEY / COLOPY STEPHEN / DANIELS DONALD
DICKERSON RICCARDO / EARNESTY JOHN / FERNANDEZ DANIEL
HARALDSON DAVID / HELTON DONALD / HILL JOE / IMAE HACHIRO
JARRELL ROGER / LOWDEN THOMAS / MORRIS ARTHUR / NORTHROP JAMES
PARNELLA JOHN / PINION DOCK / SCOTT TERRENCE / SHEEHY DAVID
SMITH GENE / SMITH OTIS / SNODGRASS GEORGE / SOLIZ ROLANDO
STARNES MILBURN / TAYLOR JIMMY / VESELY JOSEPH / WILSON CLARENCE

HQ COMPANY (24 KIA)
BROWN DONALD / CASSUBE RICHARD / CLARK LORINZER / CROWDER HERBERT
DORRIS CURTIS / ELYEA SIDNEY / ENGLISH JAMES / EPPS HERSCHEL
FAIN JAMES / GOODWIN ROBBIN / HARRIS STEVE / KASTEN DANIEL
LISZCZ ROBERT / LITTLE JOHN / NICHOLS LARRY / PALMER RONNY
PHIPPS JIMMY / SCHMID ROBERT / SHIPP KEITH / SMITH ROBERT
THOMAS JACKSON / WILHELM JOHN / WILLETT RICHARD / WORLEY ROM

Back of T-shirt that 1st Battalion, 5th Infantry 25th Infantry Division Class of 1966 wear to honor the 172 men killed in action at Cu Chi, Vietnam. (Hazel Willis Collection)

160   Chapter 8: Legacy of a Hero

Los Lunas High School JROTC color guard presenting the flags.
(Laurie Kastelic Collection)

Daniel D. Fernández: The Man Behind the Medal of Honor

Military Guard gave a 21-gun salute. (Laurie Kastelic Collection)

*Los Lunas High School JROTC performed for the veterans.
(Laurie Kastelic Collection)*

162   Chapter 8: Legacy of a Hero

Peter, Priscilla (Lollie), and James Fernández at memorial ceremony at Daniel Fernandez Park in 2008. (Valencia County News-Bulletin)

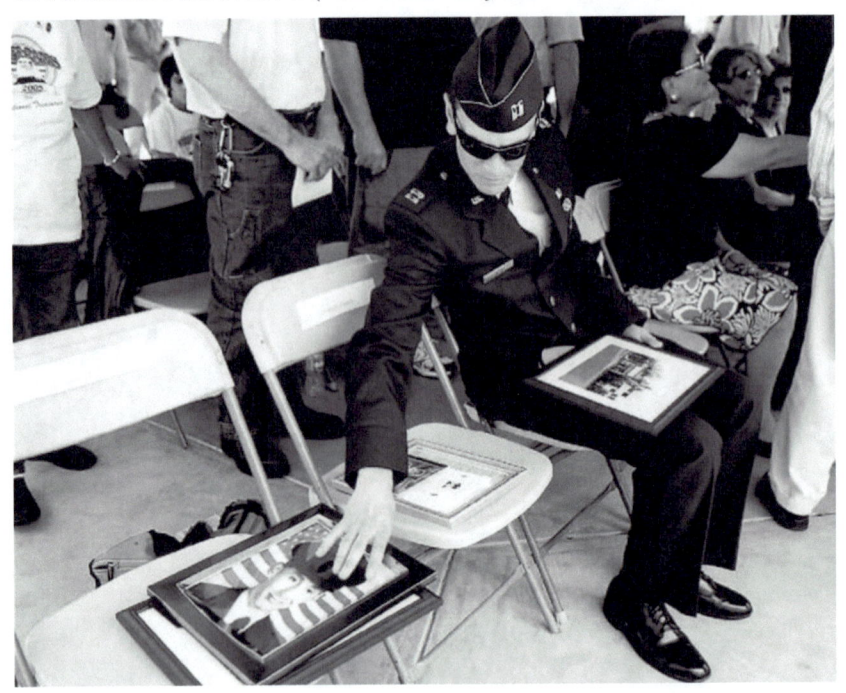

Peter and Lollie's son John Fernández touches a picture of his uncle, Daniel Fernández, during the ceremony. (Albuquerque Journal)

Los Lunas Bridge Dedication

NM Highway 6 proposed bridge replacement alignment. (Parsons Brinckerhoff)

In Fall 2015, twelve National Honor Society students from Valencia High School in Los Lunas, undertook a project to commemorate the upcoming 50th anniversary of Daniel's death in Vietnam and the 50th anniversary, in 2017, of his receiving the Medal of Honor. Project components included sponsoring an elementary school portrait drawing contest, requesting a track meet be renamed as a memorial to Daniel, sponsoring memorial bricks at the Vietnam Veterans Memorial in Angel Fire, New Mexico, and requesting legislative memorials commemorating Daniel's Medal of Honor award (2016) and all New Mexico Medal of Honor recipients (2017).

In addition, the students made an official request to have the highway bridge over the Rio Grande in Los Lunas named "The Daniel D. Fernandez Veterans Memorial Bridge ." After years of Village Council meetings, petitions, public hearings, New Mexico Department of Transportation approval, and long periods of waiting for a bridge rebuild, the Daniel D. Fernandez Veterans Memorial Bridge was completed in Spring 2021. The dedication of the bridge was held on November 5, 2021.

The dedication ceremony began with Veterans of Foreign Wars (VFW) riders crossing the bridge on their motorcycles to enter the Los Lunas River Park where the ceremony was held. Valencia County Commissioner Gerard Saiz served as the master of ceremonies, welcoming guests and giving introductions. Members of the Daniel D. Fernández VFW Post 9676 raised the colors and the Valencia High School (VHS) Honor Society lead the group in the Pledge of Allegiance.

VHS National Honor Society sponsor and teacher Laurie Kastelic told those attending the dedication ceremony about the girls that initiated the move to honor Daniel. She stated that even though the girls had graduated from Valencia High School since 2018, they stayed dedicated to honoring Daniel and stayed in touch.

"Ms. Kastelic did a really good job keeping us all together, informed. I mean, we all had different lives, obviously with college and everything," Savannah Armijo, one of the students who worked on the project and attended Friday's dedication said. "Life gets crazy, but I think she did a really good job just keeping us updated in emails and letting us know what's going on with the DOT [Department of Transportation]. With her help especially, we kept in touch with everybody."

Jordyn Peralta, one of the VHS honor students, reviewed the bridge project and Rocio Chavez thanked those that assisted in the bridge dedication process. Jordyn explained that the process was a long one and required the help and patience of many people and organizations. This included Peter and Priscilla Fernández; Daniel D. Fernández VFW Post 9676 (Chet Pino, Post Commander); James Garley (Past Commander, VFW Post 9676); Michael Jaramillo, Public Works Director for the Village of Los Lunas; Mayor Charles Griego and the Los Lunas Village Council; Stephanie Nemett, Office of Government Affairs for the New Mexico Department of Transportation; the Valencia High School National Honor Society, and the *Valencia County News-Bulletin.*

Six of the 12 VHS National Honor Society students, Jordyn Peralta, Savannah Armijo, Rocio Chavez, Courtney Eichwald, Isabel Ibarra, and Rachel Baca, who began their project to honor Medal of Honor recipient Daniel D. Fernández. (Laurie Kastelic Collection)

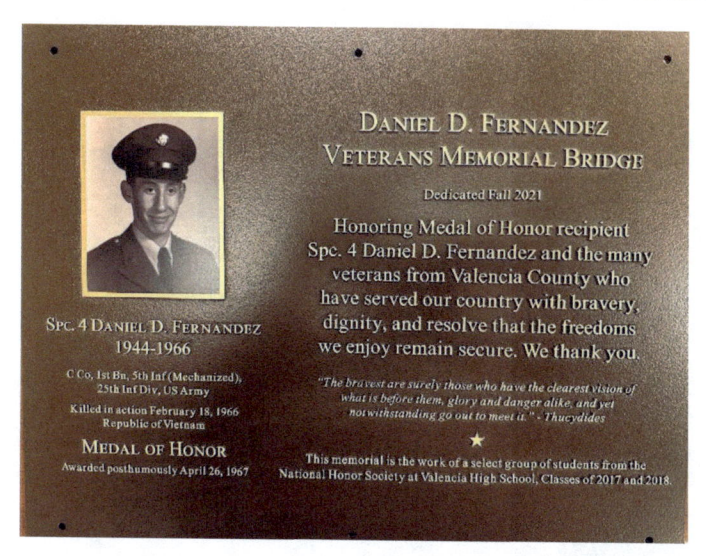

Daniel D. Fernandez Veterans Memorial Bridge Plaque. (Fernández Collection)

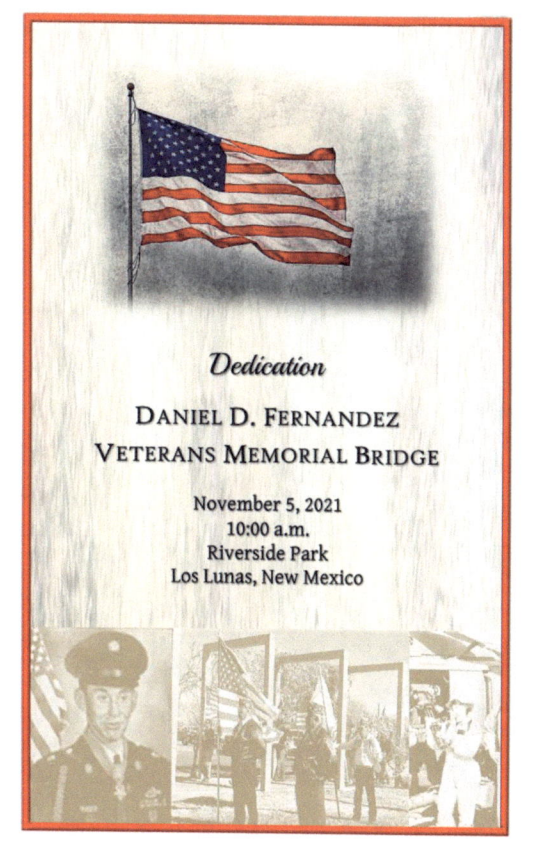

Program cover for bridge dedication. (Fernández Collection)

166   Chapter 8: Legacy of a Hero

Daniel D. Fernandez Veterans Memorial Bridge from the west side. (Cynthia J. Shetter)

Peter, Lollie, Isaac, John, and Jason Fernández at the 2021 Daniel Fernández Veterans Memorial Bridge dedication. (Fernández Collection)

"When the girls first started talking about these projects with their families at home, some of them had Vietnam veteran grandparents, who had never talked to them at all about war," Kastelic said. "But the grandparents began to tell them how important it was that they were doing this. I remember one girl telling me that her grandfather broke into tears, so it was such an important thing for her to be doing."

Another said that she looked forward to bringing her children and grandchildren to the bridge and telling them the story of Daniel and why she was so proud of what she and the others had done to dedicate the bridge to our local hero.

In 2017, Kastelic witnessed the connection grow as two of the girls traveled from Portales, where they were attending college, to the Angel Fire Vietnam Veterans Memorial for the installation of five memorial bricks the former VHS honor students had purchased in Daniel's honor with the recipient's name and Medal of Honor action date: Alexander Bonnyman, World War II November 20-22, 1943; Joe P. Martínez, World War II October 22, 1943, Harold H. Moon, World War II October 21, 1944; Alejandro R. Renteria Ruiz, World War II June 26, 1946; and Robert S. Scott, World War II July 29, 1943.

Before the gathered guests and dignitaries, the master of ceremonies introduced Courtney Eichwald and Jordyn Peralta and told the crowd, "This is our future ladies and gentlemen. I am so proud of these young ladies and Laurie Kastelic, the librarian that mentored them through this process. Thank you."

Courtney read the citation for Medal of Honor recipient Captain Florent A. Groberg who distinguished himself by stopping a suicide bomber in Asadbad, Kunar Province, Afghanistan, on August 8, 2012, during the War on Terrorism.

Jordyn read the citation for Medal of Honor recipient Joe P. Martínez who distinguished himself by rallying troops to take the strategic Holtz-Chichogof Pass position that ultimately cost him his life on May 26, 1943, during the World War II on Attu Island, Aleutian Islands, Alaska.

Courtney and Jordyn met several veterans at the event, but when the girls met Hershel "Woody" Williams, who was honored for his heroics during the Battle of Iwo Jima in World War II, the cumulative legacy of the project struck an emotional chord for project coordinator Laurie Kastelic. "I couldn't stop crying because I realized how important that really was. They were talking with a 93-year-old Medal of Honor recipient," Kastelic said. "I think it just really touched them."

168   Chapter 8: Legacy of a Hero

Eight Medal of Honor recipients are honored each year at the Vietnam Veterans Memorial in Angel Fire, New Mexico. VHS Honor students raised money to purchase five of the eight plaques laid upon the bricks. (Laurie Kastelic Collection)

Courtney Eichwald, Hershel "Woody" Williams, and Jordyn Peralta at the Angel Fire Vietnam Veterans Memorial brick installation ceremony in 2017. (Laurie Kastelic Collection)

Legacy to the Family

Daniel Fernández's death made national headlines, placing the normally reserved Fernández family into the limelight. James Fernández, Daniel's youngest brother, remembered, "At the time, my grandmother had been ill, and I was on my way home from school, and I saw all these people at our house. I just thought it was my grandmother, never thinking it would be my brother. It just seemed like an impossibility. It seemed like it was a very bad dream, and I hoped it would go away, but it never did."

James was to turn 12 years old the following week, and his birthdays would be forever marred by the memory of his brother's death despite his parents' efforts to give him normality. It was a lot for James to absorb for someone so young. He would watch his parents stoically sit through various military ceremonies that they were invited to in the years to come knowing the heartache they felt.

Rita Fernández was 19 years old and attending Highlands University in Las Vegas, New Mexico, when she received the news. Someone was sent to bring the quiet young woman home to Los Lunas.

Peter Fernández was 16 years old that year. He was a letterman at Los Lunas High School and played on the basketball team.

José and Laura kept their grief private and tried to respond to the letters that poured in from all over the nation expressing their condolences. José would handle the majority of the newspaper interviews, expressing reserved pride in his son's unselfish deed.

There is no easy formula for a family to deal with the aftermath of a death. The Fernández family has had to walk a difficult tightrope in the decades after Daniel's death. They remained buoyed by Daniel's heroism while hurt beyond belief by the finality of his loss.

"I wouldn't say we've been hardened to it," said James Fernández in a 1992 interview for the *Albuquerque Journal*. "Maybe we deal with it better because we've had to cope. I think we understand what people try to do. They aren't bringing up [Danny's death] to hurt; they are trying to feel something good."

Being public symbols of bereavement had its limitations though. On Memorial Days, family members would go to Danny's blue spruce-shaded gravesite at the Santa Fe National Cemetery early in the morning, taking care to beat the obligatory flags and bugle services that usually came later in the day. "We always leave before the ceremony," James said. "All my mother needs to do is hear '*Taps*.'"

José and James Fernández giving an Associated Press interview that ran in the Battle Creek Enquirer *in Battle Creek, Michigan, on April 10, 1966.*

Daniel's death became the Fernández family's platform to support all veterans to ensure that their memories and sacrifices are remembered. Laura became a member of Optimist International, instilling values in children that included promoting an active interest in good government and civic affairs, inspiring a respect for the law, and promoting patriotism. She was also an active member in the Gold Star Mothers, VFW Women's Auxiliary, the American Legion, and the American GI Forum.

James remembered that the family ran on military time in the 1960s and 70s. They were regulars at Memorial and Veterans Day observances throughout New Mexico and parts of Texas. They always had to take care not to show undue emotion. "That was a no-no," James said. "We were expected to show strength."

It was not unusual for the family to run into people who knew Danny or wanted to talk about him. They often received letters from some of Danny's Vietnam buddies or sometimes they would arrive on the doorstep unannounced.

José passed away in 1981 and Laura in 2006. Until their deaths, Daniel's parents continually demonstrated their support of veterans and educated their own children Rita, Peter, and James of the importance of these veterans' sacrifices.

James and Rita never married, nor did they have any children of their own. It is through Peter's family that Daniel's legacy lives on.

Priscilla "Lollie" Paiz was Peter's high school sweetheart. After they graduated from Los Lunas High School in 1967, they attended Highland University in Las Vegas, New Mexico. Peter soon realized that college was not his strong suit and decided to enlist in the Air Force. His mother tried to bribe him by offering to send him to New York to go to school. Anything to prevent the possibility of losing another son to the war.

Peter talked to the recruiter, and he was able to guarantee Peter that he would not see military action in Vietnam. This somewhat alleviated his mother's fears.

Peter was stationed at Keesler Air Force Base, Mississippi, when he and Lollie married on July 5, 1969. After a trip to Denver, they proceeded to Italy where Peter was stationed for several years.

After Peter's stint in the Air Force, they returned home to New Mexico where he received a degree from New Mexico Highlands University. The couple lived in the Denver and Littleton, Colorado, areas before moving back to Los Lunas. Peter worked as an auditor for the United State General Accounting Office (USGAO) for 25 years and the Village of Los Lunas for 15 years, having also served as the administrator for the Village.

As a schoolteacher, principal, deputy superintendent, and later the state director of AdvancED Accreditation for New Mexico Schools, Lollie has held a platform to help educate not only her and Peter's children but all children about citizenship, character, leadership, and having pride in our country's history.

Peter and Lollie have taught their children and others about Daniel's story as well as the legacy of Lollie's father, Juan Luis Paiz, who was a Bataan Death March Survivor serving in New Mexico's famed 200th Coast Artillery during World War II.

On December 8, 1941, a day after Pearl Harbor was attacked, the Japanese attacked Luzon in the Philippine Islands where Juan was stationed. He survived the initial bombardment only to surrender along with his fellow soldiers to Lieutenant General Masaharu Homma and his Japanese Army of 43,000 troops. Juan, along with an estimated

75,000 Filipino and American troops, was forced to make the arduous 65-mile march to the prison camps, enduring brutal treatment by the Japanese guards.

PVT Juan Paiz ultimately survived starvation, disease, and the brutal treatment that was rampant at Camp O'Donnell and on the POW ships that finally carried him to the labor camps in Japan. He was held there until the United States bombed Nagasaki with the Fat Man plutonium bomb on August 9, 1945, ending the war.

Juan was one of four men who chose a military career after being liberated from a Japanese labor camp at the end of World War II. "He left the Army to join the Air Force after spending 3 ½ years as a prisoner of war," said his daughter Lollie. Juan retired in October 1962, having served over 21 years in the Army and the Air Force combined.

Daniel Fernández and Juan Paiz distinguished themselves through a strength of character and resiliency fighting for freedom. Their lives contain a legacy of stories that Peter and Lollie have shared with their children Michael, Jason, and John as well as their grandchildren Camaren Ly, Jabari Carlton, Garrett, Emma, Zac, Alaina, and Isaac Fernández.

"I think the key is focusing on the importance of what he did and making [people] aware," said Peter in an interview about his brother. "Freedom is not free; it costs a lot and unfortunately it costs human lives, too. That is why we should all definitely appreciate the freedoms that we have here. I think that's what people lose sight of – it doesn't come for free."

Daniel D. Fernández: The Man Behind the Medal of Honor 173

Lollie's father, PVT Juan Luis Paiz (1919 – 1995), survived the Bataan Death March and spent 3 1/2 years in a Japanese POW camp. (Fernández Collection)

Juan Paiz's US World War II Draft Card. (Fold3.com)

Rita, John, James, and Laura Fernández at John's Air Force Academy Graduation in 2003. (Fernández Collection)

Lieutenant Colonel John Z. Fernández, Engineering Manager for the 379th Space Range Squadron at Schriever Space Force Base near Colorado Springs, holds his Uncle Daniel's Medal of Honor, 2022. (Fernández Collection)

Daniel D. Fernández: The Man Behind the Medal of Honor 175

Peter, Lollie, and Michael Fernández at 25th Infantry Division Ceremony honoring Daniel. (Fernández Collection)

Peter and Lollie's granddaughter, Camaren Ly, graduated from the United States Air Force Academy in 2020. (Fernández Collection)

176   Chapter 8: Legacy of a Hero

Alaina Fernández, Daniel's grandniece, reading the memorial plaque at Daniel Fernandez Park, 2021. (Fernández Collection)

Nine-year-old Zac Fernández composed a collaborative storyboard of his Great Uncle Daniel's life. (Fernández Collection)

Daniel D. Fernández: The Man Behind the Medal of Honor 177

The street was named Fernández Street when the Village of Los Lunas annexed the Chihuahita area that the Fernándezs' lived in. (Fernández Collection)

Emma Fernández placing American flags at the National Cemetery in Santa Fe, New Mexico, 2014. (Fernández Collection)

178   Chapter 8: Legacy of a Hero

Garrett Fernández, Daniel's grandnephew, making a crayon rubbing of Daniel's headstone at Santa Fe National Cemetery, 2014. (Fernández Collection)

Daniel Fernández's grave site in the Santa Fe National Cemetery is decorated by the family annually. (Fernández Collection)

Chapter 9: What Makes a Hero?

Daniel Fernández was a hero. There is no question about that. But what characteristics made him a hero? And which of these characteristics do we admire so much that the community of Los Lunas still honors him long after his heroic sacrifice?

There have been many studies of heroism. Each study lists characteristics that heroes consistently share. Of all these characteristics, three, in particular, appear on most lists. It is not surprising that Daniel possessed all three qualities, including empathy, conviction, and, of course, bravery. If there were a personality test to measure a person's potential for heroism, Daniel would have scored a perfect ten long before he reported for duty and combat in Vietnam.

Empathy

First, heroes display a high degree of empathy or compassionate concern for the care and safety of those around them. Beyond being sympathetic, these people are particularly able to feel what those in need are feeling.

Daniel displayed empathy in many ways, starting with his love of animals, especially horses. He loved to care for horses and was known to break them not with harsh methods to prove his bravado, but in a calm, gentle manner. And he needed horses as much as his horses needed him. They were a source of comfort and healing, as when he rode horses in Hawaii and at home while he recovered from wounds suffered during his first deployment to Vietnam. The first thing he did when he arrived in Los Lunas to recover was to buy a pair of boots and go horseback riding.

In both Vietnam and Hawaii Daniel displayed his great empathy when he intentionally sought out and befriended many native men, women, and children. He studied their languages and appreciated their cultures. He came to protect these friends because he knew them so well and admired them so much.

Daniel no doubt learned to be empathetic by watching and learning from his parents. He saw his mother and father work in community service both in Mexico and New Mexico. After Daniel's death, his mother worked for several years as the Treasurer of Valencia County. His brother Peter worked as a Federal auditor for USGAO and was the Village of Los Lunas Administrator. His sister Rita worked as a journalist in many states throughout the U.S. His youngest brother James worked as County Treasurer before helping to create and administer the Los Lunas Museum of Heritage and Arts.

The Courage of His Convictions

Usually without conceit, empathetic heroes are convinced that they have something to contribute for the betterment of others. Their conviction often leads them into service jobs, as volunteers, or in their chosen careers. They strive to help others survive and succeed with no expectation of compensation, reward, or glory.

Driven by his conviction that he should serve his country as soon as he could and as much as possible, Daniel left high school and enlisted in the Army just as the conflict in Vietnam was heating up and just before it became controversial back home. He was a child of the generation of Americans who responded to President John F. Kennedy's challenge: "Ask not what your country can do for you. Ask what you can do for your country."

A natural leader, Daniel bonded quickly with his fellow soldiers during his two tours of duty. At first, most soldiers were eager to serve in order to defend their families, their homes, and their country. But once in battle, priorities often changed, with soldiers becoming more eager to fight for their fellow soldiers as they bonded and experienced the horrors of war together. Forming tight bands of brothers, they worked as teams and made good friends with many of the men in their patrols. Everyone in the group knew what each man was capable of, and which men needed additional attention in daily survival and combat. And they instinctively knew who could be counted on the most in every circumstance.

Daniel's senses of empathy and conviction made him especially able to bond with soldiers in his company. In giving nicknames for one another, it was significant that the men in his patrol called him "Old Dan" not only because he was 21 (when most of them were 18 or 19), but also because he often looked after them like an older brother caring for his younger siblings. They always knew they could rely on Daniel, no matter what horrors they faced together.

Bravery

Of course, a hero's greatest attribute is his or her bravery at the most critical moments, even when he or she is inwardly as frightened as anyone else around them. Heroes are just better able to conceal and control their fear so that they can be of help to others. Daniel showed this ability over and over in his young life.

On his first tour in Vietnam, Daniel served in one of the army's most dangerous jobs, door gunner in helicopters as they swept into enemy zones at low altitudes to deliver or evacuate troops, including the injured. Only point men on patrol and tunnel rats who cleared out enemy tunnels had similarly perilous jobs. With only slight exaggeration, door gunners were said to have life spans of five minutes in battle. While not fatally wounded, Daniel was shot in the leg while helping three fellow soldiers under heavy fire.

After recovering from his wound, Daniel joined an infantry patrol that faced danger at every moment of every day "in country." He volunteered for his final mission, and, once in the field, volunteered to accompany his squad leader on high-risk reconnaissance missions twice within hours.

And then Daniel made his bravest decision. When the enemy had thrown a live hand grenade near him, he kicked it away, but it accidently landed near his own men. Giving no consideration to his own safety, Daniel shouted, "Move out, you people!" and dove on the grenade, covering the blast with his body. In a split second, he did exactly what he was perfectly wired to do: use his empathy, conviction, and bravery to save others by sacrificing himself.

Even as he waited to be medivacked out by helicopter, Daniel thought of others and their fate without him. Half in jest and half seriously, he asked a friend "Who's going to take care of you now?" Daniel died of his fatal wounds hours later.

Knowing what we know of him, we must wonder how Daniel might have lived with himself had he not acted as he did. His psychological agony and pain would have surpassed any physical wounds he might have suffered if he had survived.

Hometown Los Lunas

Daniel's heroism on February 18, 1966, is one of the most admirable, compelling, and tragic stories of the Vietnam era. His act of complete selflessness has been honored in many ways, especially in Los Lunas where everything from a park and a bridge to a former school and a VFW post bear his name.

Why have Daniel's hometown residents done so much to remember Daniel and his great act of courage? The answer is simple: Daniel's friends and neighbors honor not only his superior act of bravery, but also the three main characteristics that, combined, made him a hero.

We hope to exhibit these characteristics in ourselves, while teaching them to our children. It is sometimes possible to emulate these characteristics individually, but seldom in the combination that made Daniel and heroes like him so rare.

No one remembers people who refuse to make things better if the personal cost is too great. We will never forget Daniel Fernández, the man who always strove to make things better for others, regardless of the cost.

Epilogue
The Men Daniel D. Fernández Saved

On the morning of February 18, 1966, five men stood on a fine line between life and death. It was Daniel Fernández that made the final decision that ended his life but gave his comrades in arms additional time on this earth. We decided to seek out those four men he saved that day to see how they spent their additional time.

SP 4 George E. Snodgrass (1945 – 1966)

George E. Snodgrass. (Fold3.com)

George Edward Snodgrass was born March 16, 1945, in Pompton Lakes, Passaic County, New Jersey to Edward M. Snodgrass and Helen Jacobs. George spent most of his life in foster homes before enlisting in the Army at the age of 19.

After Daniel's death on February 18, 1966, the 1/5th (M) participated in a search and destroy operation. The first phase was located in an area southwest of Bao Trai near the Oriental River. The units of the battalion crossed the LD (line of departure) at 1000 hours on March 14, 1966, and encountered light enemy contact. Some armored personnel carriers (APCs) became stuck in the soft ground as they approached the river. The battalion continued the operation with a few encounters. On March 15 Charlie Company came under fire at the same time a booby trap detonated. The encounter resulted in four Bobcats wounded and one killed.

Jim Wilson had met George at basic training in Fort Dix near Trenton, New Jersey, and they became good friends while stationed in Hawaii together. When Snodgrass was shot, Wilson was there to put him on the MedEvac helicopter to get him to a hospital. "George was originally shot in the leg and it looked like he was going to live." Jim told him he would see him in a couple of days, but Snodgrass didn't make it to the hospital. While in transit, the MedEvac was shot at by the Viet Cong with ammunitions passing through the helicopter, killing Snodgrass one day before his 21st birthday.

After the war Wilson, a Monclaive, New Jersey, native, tried to locate where his friend was buried. George's foster mother was not very receptive but did tell him he was buried at Arlington. Wilson searched for years trying to find his friend's burial site to no avail. Finally, as technology advanced and information started appearing on the Internet, Wilson located Snodgrass's grave at Holy Cross Cemetery in North Arlington, New Jersey. However, the tombstone read, "Joseph Kowalski", the brother of George's foster mother. They had buried him in the same plot.

The following Memorial Day Wilson and a group of veterans dedicated George's burial site with a proper marker, and he was given a full military service.

Grave marker installed by Jim Wilson and other veterans of the Passaic, New Jersey, area. (Fold3.com)

PVT David R. Masingale (1947 – 2014)

David Ralph Masingale was born on August 20, 1947, in Encino, California, to Ralph Masingale and Betty Estill. David was 17 years old when he enlisted in July 1965. He was a graduate of West Coast Bible College near Clovis, California. After completing 11 weeks of medical training, he was deployed to Vietnam. PVT Masingale, the platoon's 18-year-old medic, was positioned behind the hut treating Joseph Benton's wounds with Daniel Fernández, SP4 James P. McKeown, SGT Ray E. Sue, and George E. Snodgrass when the Viet Cong increased their attack upon the men. Masingale treated SGT Sue's leg and Daniel Fernández after he was seriously wounded.

In an interview after Daniel's death, Masingale said of Danny "He was the kind of person you want yourself to be. You know, we'd all like to say to ourselves that we would do the same thing in the same situation. I wonder."

Masingale treated countless other soldiers before he completed his tour and returned home to Fresno, California, where he had a mishap one Saturday evening in 1968. He was driving along and had the urge to kiss his girlfriend. As he leaned over to deliver the kiss, the car jumped a curb, smashing into a fire hydrant. When he went to court, the judge told him he would do well to kiss her with his eyes open next time.

David later married, then divorced. Masingale died on Christmas Eve 2014 in Salem, Oregon. David was laid to rest with Military Honors at the Willamette National Cemetery through the Dignity Memorial Homeless Veteran Burial Program. He had no family to attend the service, so the public was encouraged to attend the military funeral services.

The Department of Housing and Urban Development estimates that 40,056 veterans are homeless on any given night. Over the course of a year, approximately twice that many experience homelessness. Only 7% of the general population can claim veteran status, but nearly 13% of the homeless adult population are veterans.

The National Coalition for Homeless Veterans states that a large number of displaced and at-risk veterans live with the lingering effects of PTSD and substance abuse, which are compounded by lack of family and social support. This was most likely the case for Masingale. One wonders how an 18 year old can wrap his mind around the events he experienced in Vietnam only to come home to anti-war demonstrations.

In a conversation about PSTD, Vietnam veteran Jesse Mancias was asked, "What made the Vietnam Conflict so different from the other wars?" He replied, "In World War I, World War II, and the Korean War soldiers did not see combat daily and they experienced events in stages." Soldiers in those wars would shoot at an enemy soldier, be shot at, see a dead body, shoot or even kill someone, and lose a comrade over a period of time that stretched into weeks if not months. Vietnam soldiers experienced all these events in a matter of minutes. They then came home to the protests. They were often accused of being "baby killers".

Veterans' organizations like the Daniel D. Fernández VFW Post 9676 make it their mission to help not only veterans of all wars cope but to educate the public of these veterans' efforts and instill a pride in country.

David R. Masingale headstone at Willamette National Cemetery in Portland, Oregon. (FindaGrave)

James P. McKeown Jr. (1944 – 2004)

James Patrick Joseph McKeown was born January 20, 1944, in Camden, New Jersey, to James P. McKeown Sr. and Gloria Cianfrani. James returned home from Vietnam and married Mary Catherine Stratton.

McKeown, known as "Mack" by his friends, went on to serve 21

Daniel D. Fernández: The Man Behind the Medal of Honor 187

years in the United States Army, finishing with a rank of Sergeant First Class (SFC). He served several tours of duty in Vietnam, where he received his aviation crewmember wings as a door gunner in a helicopter crew. He also served as a drill sergeant and instructor.

Military service built the foundation for Mack to return home to New Jersey to go into law enforcement. He served with the Camden County Sheriff's Office for 17 years where he was a member of the Special Weapons and Tactics (SWAT) team and the Color Guard. He was a well-liked officer, receiving Sheriff Officer of the Year award in 1993.

James P. McKeown (in round hat) receiving the Sheriff Officer of the Year in 1993. (Ron Karafin, Courier Post)

On the evening of February 8, 2004, Sheriff's officers were called to a hotel to investigate a disturbance between McKeown and a woman in Camden. When the officers arrived, they found McKeown in the parking lot. As they started to talk to him about the situation, he pulled a gun from the waistband of his pants and shot himself in the stomach. James P. McKeown died from the self-inflected gunshot wound.

James P. McKeown was a loving father and grandfather. His family and the Sheriff's Department were devastated. They may never know what lead him to take his life and if it had anything to do with his time in Vietnam.

John Reinmuth, president of PBA 277, which represented Camden County Sheriff's officers, was good friends with McKeown after the two attended the police academy together. He remembered him as "fair, honest, and spoke candidly. He was a good guy."

Several officers in the local communities throughout the area as well as the department received a significant portion of their training from Officer McKeown. Many remembered him with a big smile upon his face.

SFC James P. McKeown was interred at Arlington National Cemetery in Arlington, Virginia.

SGT Ray E. Sue (1942 -)

Ray Eugene Sue was born December 5, 1942, in Big Springs, Texas, to Ray T. Sue and Mary Nell Hooper.

Growing up in Texas, Ray was a Boy Scout and enjoyed sports. He volunteered with both the Police and Fire Departments and raised sheep for livestock shows.

He remembers enjoying shooting as a child but, like Danny Fernández, it earned him a sharp word from his mother. Where Danny almost put out Charles Sullivan's eye, young Ray shot his mom in the butt with a BB gun.

Ray E. Sue, 17 years old. (Ray E. Sue)

Ray went on to attend Ranger High School in Ranger, Texas, until he enlisted in the Army on December 15, 1959. He received basic training at Fort Carson, Colorado, and further training at Fort Ord near Monterrey Bay, California. Ray was sent to Korea for a 13-month stay, then Ford Hood, Texas, a year before going to Vietnam in 1964 – 65.

Ray Sue and Daniel Fernández both served in the 501st Aviation

Battalion, but Sue was with the Bearcats. He flew over 900 combat hours as a Shot Gun Rider, receiving the Air Medal and other decorations. Ray and Danny got to know each other in their off time, was mainly when the gunships were being repaired. SGT Sue saw Danny quite often with a camera in and around the Cu Chi area during the four months they were there. Although Danny's heart was with a girl back home, he would pose for pictures with young women for their unique culture and dress.

After SGT Sue's first tour to Vietnam, he was assigned to the 25th Infantry Division in Hawaii. His stay there was short lived because he and Danny were shipped out together to Vietnam for their second tour.

In an interview in 2023, Ray said they arrived in Cu Chi a week after they landed near Saigon. They had to wait for their equipment before advancing to Cu Chi. Once at Cu Chi, they had to create a perimeter and dig their foxholes. One evening he and another soldier were in the foxhole and the other soldier said, "Stop kicking me!" Sue replied, "I'm not!" This went on several times during the night. Finally, dawn broke and the soldier lifted his poncho to discover two cobras nestled next to him. "He jumped up out of that hole white as a sheet!" recalled Sue laughing. It wasn't funny at the time, but the memory brought a chuckle to the now 80-year-old veteran.

Ray E. Sue at the West Texas VA Healthcare Center in Abilene, Texas, 2022. (Vital Sign)

The retired veteran still remembers the day Danny was killed. That morning about 0735 he was on listening duty. "The first person that got hit was Joseph T. Benton. He was a machine gunner." SGT Sue got up to check on Benton, then went to check on a couple of others and told them to take cover and watch the trail. He came back and "I was looking down the line where the other people were to see if they were fine, and someone [George Snodgrass] came up behind me."

"They were having trouble," Sue said. "They were trying to carry

Benton out. So, I helped him try to get him through the trees. About that time, I got hit. Then I hit the ground and then about five minutes later, that is when Danny took the grenade out from under us, and it came back on him."

"It was one helluva fight," he said, "and I was the only one left on the field. The greatest thing I ever saw was those F4s coming in. They turned lose with a 20mm…340 rounds chewed up the ground," Sue said, "Boy, was I proud of them!" He was feeling really good sitting in his front row seat as he leaned up against a tree watching the action. "By the time I was extracted from the area, there were only two of 16 unscratched," Sue said.

The last SGT Sue saw Danny was when the medics were loading Ray on the helicopter to take him to Saigon. "I reached over," Ray paused, then continued as his voice broke with emotion, "and told him I'll see you at home." He doesn't know if Danny made it to the hospital alive. They wouldn't tell him once he awakened from surgery on his leg.

Meanwhile, SGT Sue's father, Ray T. Sue, a police officer, was used to being awakened in the middle of the night but that night a fellow officer asked him to meet him in his driveway. The officer had a newspaper clipping telling the story of Danny's death and his son Ray being wounded. This was the first that SGT Sue's parents had heard that he was wounded. They went days without knowing the condition of their son. No one would tell them anything. Finally, they received word he was at Fort Hood, Texas. He had been wounded in the leg and was peppered with shrapnel from the grenade blast.

The M60 round shattered the bone and muscle in his left thigh, resulting in several major surgeries. He spent 18 months at Darnall Army Hospital at Fort Hood. Ray's sister Marsha was 11 at the time and remembers, "His casts were from under his arms, all around his body, all the way down to his big toe on his left leg and past the knee on his right one." She laughed, remembering, "He would get tripped up and looked like a big ole tree going boom."

SGT Ray E. Sue was awarded the Bronze Star and Purple Heart for his part in the action on that day in February 1966. As Danny's commander, he recommended Danny for the Medal of Honor. Sue said it was an honor to be invited to Washington, D.C., in 1967 for Danny's Medal of Honor ceremony. While he had great men in his platoon, he wished he had more men like Danny.

SGT Sue continued in the Army as an ordinance man for three

more years. His doctor convinced him he was not suited for pushing papers behind a desk. Ray then went to college to obtain a Masters in Horticulture. He worked for the City of Fort Worth, landscaping several parks and running its greenhouses.

In his retirement, Ray E. Sue has served as a secretary and chaplain for his local VFW. He recently moved back to Ranger, Texas.

Ray still has nightmares of his time in Vietnam. He said most veterans do not talk about their experiences, but he manages and values his memories of Danny Fernández and others that served with him.

SGT Ray E. Sue (right) and a buddy. (Ray E. Sue)

Appendix 1
A Brief History of the Medal of Honor

The Medal of Honor is the nation's highest military award. Forty-nine individuals, each with a significant New Mexico connection, including Daniel Fernández, have received this most prestigious of awards, nine posthumously.

Although there are various criteria generally followed to identify a Medal of Honor recipient with a particular state, author John Taylor has chosen the inclusive set shown below:
- Born, was raised, or lived for a significant period of time in New Mexico,
- Died and/or was buried in New Mexico,
- Enlisted in the Armed Forces in New Mexico, or
- Received the Medal of Honor for direct service in New Mexico.

Taylor contends that there is no harm in sharing and recognizing these heroes as representing New Mexico, even though more restrictive criteria may assign them to another state.

The Medal of Honor was first awarded in 1862, and the first New Mexican to be given the award was Private Albert Sale who was recognized for his gallantry during a battle along the Santa Maria River in western Arizona in June 1869. His Medal of Honor was awarded on March 3, 1870, on behalf of President Ulysses S. Grant.

Since that earliest award, New Mexicans from all branches of the armed services except the Coast Guard have been recipients. In some cases, we have considerable information about their lives and the circumstances of their service. In other cases, our information is much less detailed.

Although senior military officers in Europe often received badges or awards for achieving victory, General George Washington felt that common soldiers who exhibited gallantry in battle should be similarly recognized. He stated, "The road to glory in a patriot army and a free country is...open to all." Thus, the history of awards for gallantry in the United States armed forces dates to the Revolutionary War period. The Armed Forces Decoration and Awards Publication cites the following early history:

> The first formal system for rewarding acts of individual gallantry by the nation's fighting men was established by General George

Washington on August 7, 1782. Designed to recognize "not only instances of unusual gallantry in battle, but also extraordinary fidelity and essential service in any way," the award consisted of a purple cloth heart. Records show that only three persons received the award during the Revolutionary War: Sergeant Elijah Churchill, Sergeant William Brown, and Sergeant Daniel Bissel Jr.

The Badge of Military Merit, as it was called, fell into oblivion until 1932 when General Douglas MacArthur, then Army Chief of Staff, pressed for its revival. Officially reinstituted on February 22, 1932, the now familiar Purple Heart was at first an Army award, given to those who had been wounded in World War I or who possessed a Meritorious Service Citation Certificate. In 1943, the order was amended to include personnel of the Navy, Marine Corps, and Coast Guard. Coverage was eventually extended to include all services and "any civilian national" wounded while serving with the Armed Forces.

Badge of Military Merit (Public domain)

Although the Badge of Military Merit fell into disuse after the Revolutionary War, the idea of a decoration for individual gallantry remained through the early 1800s. In 1847, after the outbreak of the Mexican American War, a "certificate of merit" was established for any soldier who distinguished himself in action. No medal went with the honor. After the Mexican American War, the award was discontinued, which meant there was no military award with which to recognize the nation's fighting men.

Early in the Civil War, a medal for individual valor was proposed to General-in-Chief of the Army Winfield Scott, but Scott felt medals smacked of European affectation and killed the idea.

However, the idea of a medal found support in the Navy, where it was felt recognition of courage in strife was needed. Congressionally approved Public Resolution 82, containing a provision for a Navy medal of valor, was signed into law by President Abraham Lincoln on December 21, 1861. The medal was "to be bestowed upon such petty officers, seamen, landsmen, and Marines as shall most distinguish themselves by their gallantry and other seamanlike qualities during the present insurrection."

Shortly after this, a resolution similar in wording was introduced in Congress on behalf of the Army. Signed into law July 12, 1862, the measure provided for awarding a medal of honor "to such non-commissioned officers and privates as shall most distinguish themselves by their gallantry in action, and other soldierlike qualities, during the present insurrection."

Although it was created for the Civil War, Congress made the Medal of Honor a permanent decoration in 1863.

The first Medal of Honor was awarded on March 25, 1863, by Secretary of War Edwin Stanton, on behalf of President Abraham Lincoln. The Medal was presented to Private (later First Lieutenant) Jacob Parrott of Company K, 33rd Ohio Volunteer Infantry, and five other soldiers for their participation in what came to be known as "The Great Locomotive Chase," a daring raid to destroy railroads and bridges in the Confederacy.

Private Jacob Parrott
(Public domain)

The Great Locomotive Chase
(Public domain)

As an interesting twist, the first engagement for which a Medal of Honor was awarded actually predated the authorization of the Medal.

In January 1894, the Medal of Honor was awarded to Brigadier General Bernard John Dowling Irwin for his action on February 13, 1861, some 17 months prior to the authorization of the Medal in July 1862. Irwin, then an assistant surgeon, led a detachment of men from the 1st Dragoons against a group of Apache led by Cochise, who had besieged 60 men from the 7th Infantry in Apache Pass in what is now southeastern Arizona Territory but what was then still part of the New Mexico Territory.

Since its establishment in 1862, there have been several designs for the Medal of Honor. The first design was used by the Army from 1862 to 1904. However, in 1869, the Grand Army of the Republic (GAR), an organization of Union Civil War veterans, designed a medal that, particularly from a distance, looked very similar to the Medal of Honor. Throughout the latter half of the 19th century, the GAR grew to become a potent political force with a membership of 410,000 by 1890. This meant that the number of GAR medals in circulation was significant. To provide more distinction between the GAR medal and the Medal of Honor, the Medal of Honor was redesigned to replace the American flag with a set of vertical stripes. In 1904, General Lewis Gillespie, Jr. (a Medal of Honor recipient himself) redesigned both the medal and the ribbon. This "Gillespie" version was used by the Army until 1944.

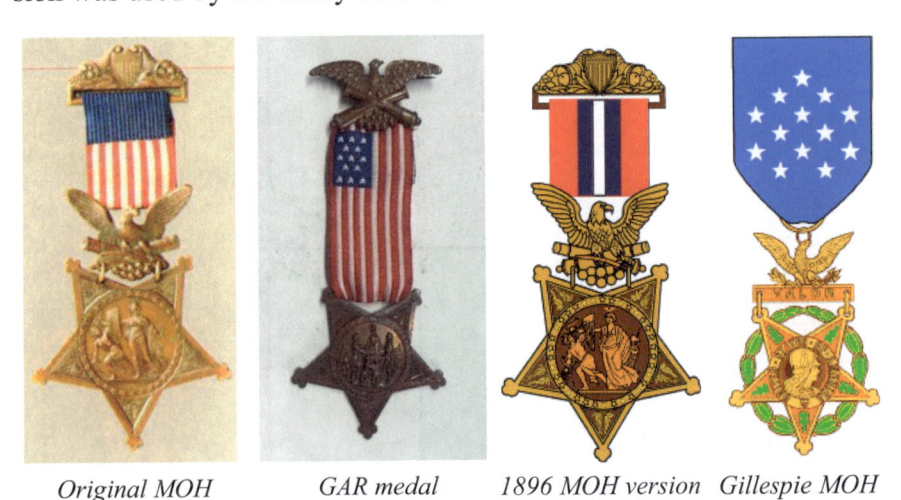

Original MOH *GAR medal* *1896 MOH version* *Gillespie MOH*

(All MOH images from Congressional Medal of Honor Society; GAR image courtesy Ken Dusenberry)

The Navy has had slightly different versions of the Medal. The version, in use from 1862 to 1912, resembled the original Army version. In 1913, the Navy adopted a different version, known as the Tiffany Cross version, designed by the Tiffany Company of New York. It was awarded to members of the Navy who earned them in combat, whereas a modified Gillespie version with an anchor replacing the eagle enabled the Navy to award Medals of Honor to men who showed extreme valor in non-combat situations, thus avoiding conflict with a 1919 law requiring combat-only awards. The Tiffany Cross version was used by the Navy from 1919 to 1942.

In 1944, the suspension medal was changed to a pendant medal to be worn around the neck. It is one of only two medals which are worn around the neck—the other is the Commander's Degree of the Legion of Merit, usually awarded to individuals who exhibit meritorious conduct while serving foreign governments.

Original Navy MOH    *Navy 1913-1942*    *Tiffany Cross (1919-1942)*

(Congressional Medal of Honor Society)

There are three modern versions of the Medal. The Army version is described by the Institute of Heraldry as… a gold five-pointed, inverted star, each point tipped with trefoils, 1½ inches wide, surrounded by a green laurel wreath and suspended from a gold bar inscribed *VALOR*, surmounted by an eagle. In the center of the star, Minerva's head surrounded by the words *UNITED STATES OF AMERICA*. On each ray of the star is a green oak leaf. On the reverse is a bar engraved *THE CONGRESS TO* with a space for engraving the name of the recipient. The pendant and suspension bar are made of gilding metal, with the eye, jump rings, and suspension ring made of red brass. The finish on the pendant and suspension bar is hard enameled, gold plated and rose gold plated, with polished highlights.

The Navy and Air Force versions are slightly different. Note that Marine Corps recipients are given the Navy Medal of Honor since the Marine Corps is a part of the Department of the Navy. The sole Coast Guard awardee, Signalman First Class Douglas A. Munro, posthumously received the Navy version of the medal for his heroism on September 27, 1942, during the battle for Guadalcanal.

Present-day Army   *Present-day Navy*   *Present-day Air Force*

(*Congressional Medal of Honor Society*)

Although the Medal of Honor is only worn as a pendant, there is a ribbon which is worn by recipients on their normal uniform.

Official Medal of Honor service ribbon.
(en.m.wikipedia.org)

To date, there have been over 3,500 awards of the Medal of Honor, over 600 of which have been awarded posthumously. These have been awarded for gallantry in all 22 of our country's conflicts from the Civil War through the Global War on Terrorism.

Only one Medal of Honor has been awarded to a woman. Dr. Mary Edwards Walker (1832-1919), a suffragist, spy, prisoner of war, and surgeon, was awarded the Medal for "Meritorious Service" during the

Civil War. Although her Medal was stripped from her and several others between 1916 and 1917, it was reinstated in 1977.

1,520 Medals of Honor were awarded to Civil War veterans, and 417 to veterans of the Indian campaigns. The last Civil War medal was granted in 1917. By 1876, a review board had decided the medal should only be given to those whose heroism was above and beyond the usual standard of duty.

Dr. Mary Edwards Walker wearing her Medal of Honor. (Public domain)

Appendix 2

New Mexico's Recipients of the Medal of Honor
(Posthumous awards are emboldened)

1st Lieutenant Alexander "Sandy" Bonnyman

Sergeant Alonzo Bowman

Sergeant Thomas Boyne

Corporal Frank Bratling

2nd Lieutenant George Burnett

Seaman Watertender Edward A. Clary

Lieutenant Matthias Day

Sergeant John Denny

Staff Sergeant Drew Dix

Private Edwin Elwood

2nd Lieutenant Robert Emmet

Specialist 4th Class Daniel D. Fernández

Corporal Clinton Greaves

Corporal Jacob Guenther

Staff Sergeant Ambrosio Guillen

Assistant Surgeon Bernard J. D. Irwin

Staff Sergeant Delbert O. Jennings

Sergeant George Jordan

Sergeant Leonidas S. Lytle

Private Joe P. Martínez

1st Lieutenant Robert McDonald

1st Sergeant James McNally

Private Charles H. McVeagh

Staff Sergeant Franklin Miller

Corporal Hiroshi Miyamura

Private Harold Moon

1st Sergeant James L. Morris

Captain Raymond G. Murphy

Corporal Thomas Murphy

1st Sergeant Francis Oliver

Staff Sergeant Leroy Petry

Sergeant First Class Louis Rocco

Sergeant Yuma William Rowdy

Private 1st Class Alejandro Ruiz

Private Albert Sale

Wagoner John P. Schnitzer

1st Lieutenant Robert S. Scott

Corporal Edward Clay Sharpless

Sergeant Thomas Shaw

Blacksmith John Sheerin

Private Eben Stanley

Private 1st Class José Valdez

Brigadier General Kenneth Walker

Private Augustus Walley

1st Lieutenant Wilbur Wilder

1st Sergeant Moses Williams

Private Henry Wills

Sergeant Brent Woods

Lance Corporal Kenneth Worley

Appendix 3

New Mexico Medal of Honor Recipients Buried at Santa Fe National Cemetery with Daniel Fernández

Edwin L. Elwood
(Indian Wars, 10/20/1869)

Jacob Guenther
(Indian Wars, 1868 - 1869)

Thomas Murphy
(Indian Wars, 8/25/1869)

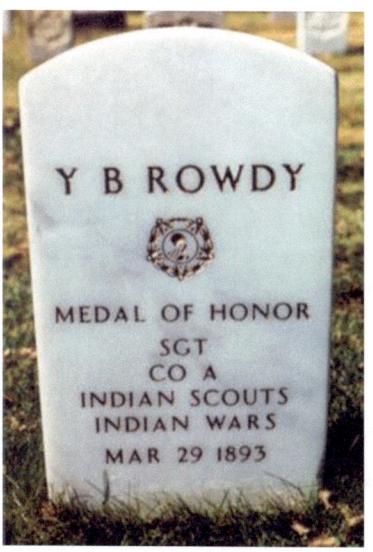

Native American Scout Y.B. Rowdy (Indian Wars 3/7/1890)

Images: FindaGrave.com

Appendix 3

Edward Alvin Clary
(USS Hopkins, 2/14/1910)

Robert Sheldon Scott
(World War II, 7/29/1943)

José F. Valdez
(World War II, 2/17/1945)

Raymond Gerald Murphy
(Korea, 2/3/1953)

Images: (FindaGrave.com)

Appendix 4: Timeline

June 30, 1944	Daniel Damaso Fernández born in Albuquerque, New Mexico.
1955	Fernández family moves to Los Lunas.
1959	First American military advisors deployed to Vietnam.
1962	Daniel drops out of high school and enlists in the Army. Boot camp at Fort Polk, Louisiana.
November 27, 1964	Daniel arrives in Bien Hoa, Vietnam, as a part of the 25th Infantry Division.
February 24, 1965	Daniel assigned to "fly shotgun" with 501st Aviation Battalion.
March 2, 1965	Daniel is shot in the leg during a helicopter operation.
March 17, 1965	Daniel is sent to Hawaii for rest and recuperation (R & R) from his injuries; awarded the Air Medal for participating in 25 helicopter missions. His R&R includes a 30-day leave in Los Lunas.
March 19, 1965	Daniel is awarded the two Purple Hearts in recognition of his injuries.
December 1965	Daniel is redeployed to Vietnam.
January 1966	Daniel's unit moved to camp at Cu Chi.
February 18, 1966, 0100	Daniel's squad leaves base camp on a reconnaissance patrol.

Appendix 4

February 18, 1966, 0700	Daniel's squad comes under fire from Viet Cong.
February 18, 1966, 0700	Daniel jumps onto a grenade to protect his squad members. His injuries are life-threatening, so he is medevacked to the Army Hospital.
February 18, 1966, ~1000	**Surgeons are unable to stop internal bleeding. Daniel dies at the Army Hospital.**
February 18, 1966, 1441	Daniel's parents receive the telegram notifying them of Daniel's death.
February 21, 1966	Daniel Fernández's company officers nominate him for the Medal of Honor.
February 25, 1966	Daniel's body arrives in Belen on the San Francisco Chief after being flown from Vietnam to Travis Air Force Base in California.
February 26, 1966	Mayor Howard Simpson declares February 26 as Daniel Fernández Day in Los Lunas. All flags to be flown at half-staff.
February 26, 1966, 2000	Rosary recited at San Clemente Catholic Church after Daniel's body has lain in state.
February 27, 1966, 0830	Father Francis Schuler celebrates a High Requiem Mass in the Los Lunas High School gymnasium.
February 27, 1966, PM	**Daniel is laid to rest in the Santa Fe National Cemetery with full military honors.**
February 1966	Roberto Martínez writes *"El Corrido de Daniel Fernández."*

April 1966	New middle school in Los Lunas named for Daniel Fernández.
May 18, 1966	VFW Post 9676 renamed for Daniel Fernández.
October 1966	Secretary of Defense Robert McNamara approves the recommendation for the award of the Medal of Honor, and forwards his recommendation to the Joint Chiefs of Staff.
November 30, 1966	Gymnasium/auditorium at Los Lunas Hospital and Training School dedicated to Daniel Fernández.
February 1967	New Mexico Senate passed Joint Memorial honoring Daniel Fernández.
April 6, 1967, 1300	**President Lyndon B. Johnson presents the Medal of Honor to Daniel's parents at the White House in Washington, D. C.**
June 29, 1967	Daniel Fernández Park in Los Lunas dedicated by Senator Clinton P. Anderson.
1972	Dedication of Daniel Fernández Recreation Center in the Daniel Fernández Park.
September 14, 1974	Obelisk installed in Medal of Honor Grove in Valley Forge, Pennsylvania.
June 19, 1975	Fernandez Hall at Schofield Barracks in Hawaii dedicated in Daniel's name.
October 25, 1976	Plaque placed at VA Hospital in Albuquerque.

November 13, 1982	Vietnam Veterans Memorial Wall in Washington, D.C., dedicated with Daniel's name in panel 5E, line 46
October 17, 1984	Fernandez Hall on Kirtland Air Force Base dedicated.
November 4, 2005	Wall of Honor dedicated at the New Mexico State Capital.
2009	Book on Daniel Fernández published by students at Daniel Fernandez Intermediate School.
February 16, 2016	New Mexico House of Representatives declares Daniel Fernández Day.
October 6, 2017	Rededication of Fernandez Hall at Kirkland Air Force Base.
November 8, 2018	Exhibit of New Mexico's Hispanic Medal of Honor recipients opens at National Hispanic Cultural Center.
November 5, 2021	New bridge across the Rio Grande in Los Lunas named the Daniel D. Fernandez Veterans Memorial Bridge.

Appendix 5

Valencia County and Isleta Pueblo Residents Killed During the Vietnam Conflict

Four hundred fifty New Mexicans were killed during the Vietnam Conflict. These are the men from Valencia County and Isleta Pueblo on that sacred list.

Secundino Baldonado, United States Air Force, May 16, 1965

Robert David, United States Army, May 21, 1968

Daniel D. Fernández, United States Army, February 18, 1966

Eddie Leonard García, United States Army, June 30, 1969

Luperto García, United States Marine Corps, June 27, 1968

José Bernardino Gonzales, United States Marine Corps, December 9, 1968

Michael Leon Lovato, United States Marine Corps, February 9, 1968

Eddie Anthony Martínez Jr, United States Army, February 10, 1969

Hilário Moreno, United States Marine Corps, November 23, 1968

Mark Alvan Tafoya, United States Marine Corps, October 2, 1970

John Stanton Wilson, United States Army, February 23, 1969

Bibliography

Albright, Paul. "His Hometown Will Never Forget Danny Fernández." *Battle Creek Enquirer and News,* 10 Apr. 1966, p. 8.

Albuquerque Journal: various articles.

Almodova, Ramon W. and Rogers, J. David. *The Tunnels of Cu Chi. Sappers Eagle* Power Point Slide Set. https://web.mst.edu/rogersda/umrcourses/ge342/Cu%20Chi%20Tunnels-revised.pdf. Accessed 12 Jan. 2023.

Apodaca, Frida Zuni. Interview by John Taylor, 24 July 2022.

A Veteran Named Sue. *Vital Signs.* West Texas VA Heath Care Center, Fall 2022.

Award Fernández 3rd Purple Heart. *Valencia County News,* 5 Apr. 1966.

Barnes, Marsha Sue. Interview by Cynthia J. Shetter, 4 February 2023.

Beasley, Herbert. *Operation Shotgun.* Centaurs in Vietnam 1961 – 1973. www.centaursinvietnam.org/WarStories/WarDiscussions/D_OperationShotgun.html. Accessed 21 Jan. 2023.

Bobcats 1966: The First Battalion (Mechanized Fifth Infantry Twenty-Fifth Infantry Division in the Vietnam War 1966 – 1971. (n.d.). 1st Bn(M) 5th Infantry Society of Vietnam Combat Veterans, Inc. Accessed 8 February 2023, from www.bobcat.ws/history1966.html.

Breedlove, SSG Howard C. *Aerial view of 25th Infantry Division Headquarters at Base Camp near Cu Chi,* US Army. (27 Mar.1966.

Boeck, Jim. "The Last Full Measure of Devotion: Daniel Fernández in Vietnam." A *River Runs Through Us.* Ed. Richard Melzer and John Taylor. Albuquerque, Rio Grande Books, 2015.

Carland, John M. *Combat Operations: Stemming the Tide,* May 1965 to October 1966. Washington DC: Center of Military History U.S. Army, 2000.

Castillo, Cecelia Otero. Interview by Cynthia J. Shetter, 31 January 2023.

Casaus, Phill. "Echoes of Heroism," *Albuquerque Journal,* 11 November 1992, p. A1, A9.

Daniel D. Fernández Collection, 1944-2022. Los Lunas Museum of Heritage and Arts, Los Lunas, New Mexico.

"Ex-Ciscoan with Viet Army Hero," *Abilene Reporter News,* 6 March 1966, p. 3.

Fernández Collection, 1966-2023. Los Lunas Museum of Heritage and Arts, Los Lunas, New Mexico.

Fernández, Daniel Damaso. Sacred Heart Catholic Church Baptismal Records. Albuquerque, NM. Baptized 10 September 1944.

"Fernández, José and Laura." *Rio Abajo Heritage.* Belen: Valencia County Historical Society, 1981.

Fernández, James. Interview by Patricia Guggino, Los Lunas Museum of Heritage & Arts, 2006.

Fernández, Peter and Priscilla. Interviews by Richard Melzer, Cynthia J. Shetter, and John Taylor. 2022-2023.

Fox, Debra. "National Assessment Group Honors War Hero," *Valencia County News-Bulletin.* 1 November 2017.

García, Clara. "Comrades-in-arms pay loving tribute to Fernández," *Valencia County News-Bulletin,* 16 July 2008.

Goble, W. Kent. "A definite reactionary," *The Magna Times,* 10 November 1994, p. 5.

Goodrich, Thomas. *Notable New Mexican: Lindy Blaskey.* New Mexico Music Commission. 22 October 2019. www.newmexicomusic.org/2019/10/22/blaskey-lindy/ Accessed 31 Jan. 2023.

Gurulé, Frank. Interviews by Richard Melzer and John Taylor, 24 July 2022, and 19 August 2022.

Harmer, David. (2022, June 14). *What, to him, was the American flag?* Freedoms Foundation at Valley Forge. www.freedomsfoundation.org/2022/06/14/what-to-him-was-the-american-flag/. Accessed 17 January 2023.

JFK and Vietnam. John F. Kennedy Presidential Library and Museum. www.jfklibrary.org/learn/education/teachers/curricular-re-sources/jfk-and-vietnam-the-september-1963-tv-interviews. Accessed 23 January 2023.

Jones, Bob. "Shotgunners Training for Viet," *The Honolulu Advertiser,* 1 May 1965, pp 1, 4.

Jones, Bob. "Schofield Shotgun Project Stopped," *The Honolulu Advertiser*, 12 November 1965, pp 1, 8.

Jones, Thomas. *Chapters from Tropic Lightning History: 25th Infantry Division "Shotgun" Vietnam.* 25th Infantry Division Association. www.1-14th.com/Vietnam/Misc/Operation_Shotgun.html. Accessed 21 January 2023.

Karnow, Stanley. *Vietnam—A History.* New York: Viking Press, 1983.

Kastelic, Laurie. Interview by Richard Melzer, Cynthia J. Shetter, and John Taylor, 2022.

La Luna. Los Lunas High School, 1963.

Larry Artiaga Collection, 1966-2022. Los Lunas Museum of Heritage and Arts, Los Lunas, New Mexico.

Laurie Kastelic Collection, 1966-2022. Los Lunas Museum of Heritage and Arts, Los Lunas, New Mexico.

Lopez, Angie S. *Blessed Are the Soldiers.* Albuquerque: Sandia Publishing Co., 1990.

Mancias, Jesse. Interview by Cynthia J. Shetter, 1999.

Mangold, Tom, and John Penycate. *The Tunnels of Cu Chi: A Harrowing Account of America's "Tunnel Rats" in the Underground Battlefields of Vietnam.* New York: Ballatine Books, Presidio Press, 2005.

Medal of Honor Museum. Patriots Point Naval & Maritime Museum. 2020. www.patriotspoint.org/explore/medal-of-honor-museum/. Accessed 7 February 2023.

Melzer, Richard, et al. *A History of New Mexico Since Statehood.* Albuquerque: University of New Mexico Press, 2011.

Melzer, Richard. *Sanatoriums of New Mexico.* Arcadia Publishing, 2014.

"Mineral Wells GI Will Attend Ceremony for Fellow Soldier," *Fort-Worth Telegram Star*, 5 April 1967, p 5C.

Mirabal, R. S. Interview by Richard Melzer, 24 Jul. 2022.

Munoz, Cruz. Interview by Richard Melzer, 24 Jul. 2022.

Munoz, Paul J. Interview by Richard Melzer, 24 Jul. 2022.

Veteran Homelessness. National Coalition of Homeless Veterans. nchs.org/veteran-homelessness. Accessed 3 Feb. 2023.

NM 6 Bridge Replacement. Parsons Brinckerhoff. Public Meeting, 11 April 2017, CN: A300423.

New York Times News Service. "He Laid Down His Life for His Comrades," *The Fresno Bee,* 21 February 1966, p. 1, 4.

Norman S. Petit: Soldier's Medal. The Hall of Valor Project. valor.militarytimes.com/hero/500823. Accessed 10 Feb. 2023.

Nott, Robert. "Vietnam: The next forgotten war?" *Santa Fe New Mexican*, 11 November 2019, p. A1, A8.

Out Here in Viet – Nam by Lindy Blaskey. *YouTube.* Vietnam War Song Project. 18 September 2014. www.youtube.com/watch?v=UasF-Fz5gjI. Accessed 17 Feb 2023.

Patterson, Michael Robert. "Felix Z. Longoria – Private, United States Army." *Arlington National Cemetery.* 17 January 2023, www.arlingtoncemetery.net/longoria.htm. Accessed 17 February 2023.

Patterson, Michael R. "James P. McKeown-Sergeant, United States Army." *Arlington National Cemetery.* 17 Jan 2023, www.arlingtoncemetery.net/jpmckeown.htm. Accessed 17 Feb. 2023.

Petit, Norman S. "Butch". *In Memory of Armando Tesillo and Danny Shearin Killed in Action on the 31st of January 1966.* Facebook. 30 January 2021, www.facebook.com/nam66tunnelrat25th. Accessed 17 February 2023.

Prominent Hispanic Americans. Arlington National Cemetery. www.arlingtoncemetery.mil/Explore/Notable-Graves/Hispanic-Americans. Accessed 12 January 2023.

Representative Inouye (HI). *The 25th Infantry Division, Shotgunner.* Congressional Record 23 June 1965, p. 14544-14545. GovInfo. www.govinfo.gov/content/pkg/GPO-CRECB-1965-pt11/pdf/GPO-CRECB-1965-pt11-2-2.pdf. Accessed 2 February 2023.

Richins-Varela, Rebecca. Interview by Richard Melzer and Cynthia J. Shetter, July 2022.

Román, Iván. "When a Fallen Mexican American War Hero Was Denied a Wake, a Civil Rights Push Began." *History Channel.* 9 October 2020, www.history.com/news/mexican-american-rights-longoria-lyndon-johnson-hector-garcia. Accessed 12 Jan. 2023.

Senator Montoya (NM). *Presentation of the Medal of Honor by the President of the United States to the Family of SP4C Daniel Fernández, U.S. Army, of Los Lunas, NM.* Congressional Record, 6 Apr. 1967, p. 8552-8553. GovInfo. www.govinfo.gov/content/pkg/GPO-CRECB-1967-pt7/pdf/GPO-CRECB-1967-pt7-2-1.pdf. Accessed 2 Feb. 2023.

Shearin, Dan Rogers, SGT. Fold3.com by Ancestry. www.fold3.com/memorial/632032004. Accessed 21 January 2023.

Shetter, Cynthia. (2022). *Valencia County Genealogy.* Ancestry.com.

Shot Gun Platoon: How the "Shot Gun" Platoon Began. *118th Assault Helicopter Company: Thunderbirds.* www.118ahc.org/shot%20gun%20platoon.htm. Accessed 21 January 2023.

Students of the Daniel Fernandez Memorial Center Project. *Man of Honor: The Story of Daniel D. Fernández.* Ed. Laurie Kastelic. Bloomington, Indiana: Author House Publishing, 2009.

Sue, Ray E. Interview by Cynthia J. Shetter, 4 February 2023.

Sullivan, Charles. Interview by Richard Melzer and Cynthia J. Shetter, July 2022.

Sullivan, SP5 Lawrence J. *Personnel of the 2nd Brigade, 25th Infantry arrive at staging area near Ben Hoa, Vietnam.* U.S. Army, 18 January 1966.

Sullivan, SP5 Lawrence J. *Troops of the 2nd Brigade, 25th Infantry Division set up tents near Ben Hoa, Vietnam.* U.S. Army, 18 January 1966.

Tropic Lightning: A History of the 25th Infantry Division. Information Office, 25th Infantry Division, 1970. www.i-kirk.info/2nd14th/Vietnam/Archives/Publications/Tropic%20Lightning/1968%20Tropic%20Lightning%20Booklet%20d8574.pdf. Accessed 15 January 2023.

Highest Honor to Fernández. *Tropic Lightning News.* 25 November 1966. www.25thida.org/TLN/tln1-40.htm. Accessed 18 January 2023.

U.S. Army 25th Infantry Division. *The 25th's 25th…in Combat: Tropic Lightning 1 Oct 1941 – 1 Oct 1966.* Doraville, Georgia: Albert Love Enterprises, Inc., 1966.

U.S. Army 25th Infantry Division in Vietnam 1966-68 "Ready to Strike" (1968). *YouTube,* uploaded by PeriscopeFilm, 28 August 2018, www.youtube.com/watch?v=OzfcBAHCUIk. Accessed 2 February 2023.

"Vietnam: Its customs and traditions," *Tropic Lightning News:* Orientation Edition. U.S. Army 25th Infantry Division, 1967. www.25thida.org/TLN/Orientation.htm Accessed 21 January 2023.

Walker, Rose Marie. "Song Dedicated to Memory of Hero Daniel Fernández," *Albuquerque Tribune,* 10 April 1966, pp. 30.

Weekly Compilation of Presidential Documents (Weekly Comp. Pres. Docs.) 10 April 1967, pp. 598 - 99.

West, Henry Lee "Butch". Interview by Richard Melzer, 24 July 2022.

White House Memorial Album, 1967. Daniel D. Fernández Collection. 1966-2022. Los Lunas Museum of Heritage and Arts, Los Lunas, New Mexico.

Wilham, T.J. "Veterans Reunite in Honor of Hero," *Albuquerque Journal,* 13 July 2008, p. 1-2.

Willis, Hazel Shearin. Daniel D. Fernández Collection. Los Lunas Museum of Heritage & Arts. Physical and Digital Collection.

Willis, Hazel Shearin. Interview by Cynthia Shetter, 21 January 2023.

Index

A

Aguirre, Specialist First Class Anthony, 71, 85, 91,
Anderson, Senator Clinton P., 85, 120, 207
Andress, Captain James G., 85
Aragon, Vanessa, 146
Archibeque, Miguel, 105-106
Armijo, Savannah, 164
Artiaga, Larry, 147
Artiaga, Salo, 11

B

Baca, Rachel, 164
Baca, Walter, 11
Baldonado, State Representative Alonzo, 113
Baldonado, Secundino, 209
Balido, Nick, 11
Becker, Jim, 18
Benton, Specialist 4th Class Joseph T., 62-63, 185, 189-190
Bissel, Sergeant Daniel, 194
Blaschke, Linden, 109-111
Blaskey, Lindy (see Blaschke, Linden)
Bonnyman, Lieutenant Alexander, 114, 137, 142, 167, 201
Bouts, Mrs., 150
Bowman, Sergeant Alonzo, 201
Boyne, Sergeant Thomas, 201
Bradley, General Omar, 136
Bradford, State Police Chief John, 74
Bratling, Corporal Frank, 201
Brodbeck, Colonel William D., 54, 61
Brown, Brigadier General Burton R., 74
Brown, Sergeant William, 194
Bryant, Anita, 106
Burnett, George, 201

C

Campbell, Governor Jack, 119
Carrasco, Marissa, 146
Castillo, Fermin, 133
Castillo, Randy, 11
Castillo, Cecilia Otero, 69
Chavez, Rocio, 164
Chavez-Simpson, Gloria, 133
Churchill, Sergeant Elijah, 194
Cianfrani, Gloria, 186
Clary, Watertender Edward A. 201, 204
Cochise, 196
Cunningham, Lieutenant Colonel Robert K., 21
Curley, Ashley, 146

D

David, Robert, 209
Day, Lieutenant Matthias, 201
Denny, Sergeant John, 201
Dix, Staff Sergeant Drew, 114, 201
Drummond, Heather, 146
Duncan, Terry, 117

E

Eichwald, Courtney, 164, 167-168
Eisenhower, General Dwight D., 19, 136
Elwood, Private Edwin, 201, 203
Emmet, Lieutenant Robert, 201
Estill, Betty, 185

F

Fajardo, State Representative Kelly, 113
Feliciano, José, 106
Fernández, Alaina, 172, 176
Fernández, Emma, 134, 172, 177
Fernández, Frank David, 7
Fernández, Garrett, 134, 172, 178
Fernández, Isaac, 166, 172
Fernández, James Henry, 9, 37, 74, 76, 85, 91, 95, 99, 101, 106, 129, 153, 156, 162, 169-171, 174, 180
Fernández, Jason, 129, 166, 172
Fernández, Joseph Anthony, 7
Fernández, Lieutenant Colonel John, 2, 129, 162, 172, 174
Fernández, Lollie (see Priscilla Fernández)
Fernández, Lorinda (Laura), 7-9, 69-70, 74, 76, 79, 81, 85, 90-92, 94-96, 98-100, 112, 119, 129, 169-171, 174
Fernández, Louis, 9-10
Fernández, Michael, 175
Fernández, Peter, 8, 11, 74, 76, 85, 91, 95, 99, 101, 113, 120, 123, 130, 132-133, 146-148, 150, 162, 164, 166, 169, 171-172, 175, 180
Fernández, Priscilla "Lollie", 101, 113, 120, 125, 129, 132-133, 150, 162, 164, 166, 171-173, 175
Fernández, Rita, 8-9, 12, 74, 76, 85, 91, 95, 99, 169, 171, 174, 180
Fernández, Zachary, 135, 172, 176
Flores, Ray, 105

G

Gabaldon, Raymond, 1
García, Eddie Leonard, 209
García, Luperto, 209
García, John, 155-156
Garley, James, 157, 164

Gillespie, General Louis, 196-197
Glass, Baylee, 150
Gómez, Tony, 11
Gómez, Willie, 11
Gonzales, Jose Bernardino, 209
Gonzales, Rafael E., 143
Grant, President Ulysses S., 193
Greaves, Corporal Clinton, 201
Green, Joe, 109
Greenhouse, Paul S., 130-131
Griego, Angelita Telles, 7
Griego, Elijah, 133
Griego, Mayor Charles, 164
Griego, Katarina, 133
Groberg, Captain Florent A., 167
Guenther, Corporal Jacob, 225
Gurulé, Frank, 11, 81
Gurulé, Nicole, 146

H

Henderson, Ramona, 21
Ho Chi Minh, 3-4
Homma, Lieutenant General Masaharu (Japanese Army), 171
Hooper, Mary Nell, 81, 188
Horner, Major General John P., 130, 132
Howard, Brenda, 133
Hurley, Major General Patrick, 80
Hurricane, Al (see Alberto Sanchez)

I

Ibarra, Isabel, 164
Inouye, Senator Daniel K. (Hawaii), 24
Irwin, Brigadier General Bernard D. J., 196, 201

J

Jacobs, Helen, 183
James, Harry, 120
Jaramillo, Michael, 164

Jennings, Staff Sergeant Delbert, 137, 201
Johnson, Claudia Alta "Lady Bird", 89, 94
Johnson, Colonel Lynnwood M. Jr., 54, 56, 61, 85, 99
Johnson, President Lyndon B., 4, 85-87, 89, 92, 94-97, 207
Johnson, Rear Admiral Ralph C., 69
Jolly, Major General John P., 74
Jones, Bob, 23-24
Jordan, Sergeant George, 201

K

Kastelic, Laurie, 113, 146, 151, 154, 157-158, 163-164, 167
Kasky, Katie, 150
Kennedy, President John F., 4, 73, 100, 145-146, 180
King, First Lady Alice, 137
King, Governor Bruce, 139
Kowalski, Joseph, 184

L

Lin, Maya, 133
Lincoln, President Abraham, 195
Liverton, Lynn Weiler, 124
Longoria, Private Felix Z., 89
Lopez, Angie, author, 70
Lovato, Michael Leon, 20
Lujan, United States Representative Manuel Jr., 128
Luna, Fred, 1
Luna, Solomon, 1,
Luna Mansion, 10
Lutz, Rebecca, 13, 15
Ly, Camaren, 123, 126, 172, 175
Lytle, Sergeant Leonidas S., 201

M

MacArthur, General Douglas, 194
MacLeish, Archibald, 155-156
Maestas, Claire, 133
Maestas, Mike, 133
Makris, Master Sergeant Tiffany, 130
Marquez, Rosalie, 12
Martínez, Eddie Anthony, 209
Martínez, Private Joe P., 114, 137, 144, 167, 201
Martínez, Ramona, 107
Martínez, Roberto, 105-109, 206
Martínez, State Representative Ken, 113
Martínez, Top, 154
Masingale, Private First Class David R., 63, 185-186
Masingale, Ralph, 185
Maximos, 102-130
McDonald, Lieutenant Robert, 142, 201
McNally, Sergeant James, 201
McKeown, James P. Sr, 186
McKeown, Sergeant First Class James P., 63, 68, 97, 99, 185-188
McNamara, Secretary of Defense Robert, 85, 207
McQueen, State Representative Matthew, 113
McVeagh, Private Charles H., 201
Melzer, Richard, 145
Mendoza, Veronica, 146
Miles, Adjutant General Major General Frank E., 139
Miller, Sarah, 146
Miller, Staff Sergeant Franklin D., 114, 137, 201
Miyamura, Corporal Hiroshi, 114, 137, 142, 201
Monavica, Fermin, 133
Montoya, Della (Mrs. Joseph Montoya), 74
Montoya, Senator Joseph P., 74, 80-81, 104-105
Moon, Private Harold H., Jr., 114, 137, 167, 201

Moreno, Hilario, 209
Morris, Sergeant James, 201
Munoz, Cruz, 11
Munro, Signalman First Class Douglas A., 198
Murphy, Captain Raymond G., 114, 142, 201, 204
Murphy, Corporal Thomas, 201, 203

N

Nemett, Stephanie, 164
Ngo Dinh Diem, 4

O

Ohlenburger, Captain Cliff, 21
Oliver, First Sergeant Francis, 137, 201
Olson, Major General Eric T., 122
Oroña, Joseph, 150
Ortiz, Elsie, 85, 91
Otero, State Police Officer Manuel, 74

P

Paiz, Juan Luis, 171-173
Parrott, Private Jacob, 195
Patterson, Staff Sergeant Dennis, 80
Payne, Sergeant Ronald H., 57
Peralta, Jordyn, 164, 167-168
Perkins, Sergeant Ruben, 64
Petit, Sergeant Norman, 56, 158
Petry, Staff Sergeant Leroy, 114, 144, 201
Pino, Chet, 156, 164
Pino, Pete, 81
Piro, Lydia, 133
Pruitt, Holly, 156
Pullins, Lance Corporal Lee, 71

R

Rainaldi, State Senator Lidio G., 114
Reeves, Colonel John M., 128-129
Reinmuth, John, 188

Resor, Secretary of the Army Stanley R., 85, 87-88
Richins, Rebecca, 10, 13, 81
Rigdon. Grady, 133
Riley, Dana, 155
Rocco, Sergeant Louis R., 114, 144, 201
Romero, County Appraiser M.S., 11
Romero, Technical Sergeant Foch, 1
Rowdy, Sergeant of Scouts Yuma William, 201, 203
Ruiz, Private First Class Alejandro R., 114, 142, 144, 167, 201

S

Saiz, Gerard, 163
Sale, Private Albert, 193, 201
Sanchez, Alberto, aka Al Hurricane, 105
Sanchez, State Senator Clemente, 113
Sanchez, State Senator Michael, 113
Schnitzer, Wagoner John P., 201
Schuler, Father Francis, 70, 73-74, 78, 85, 91, 99, 206
Schwartz, Robert, 81
Scott, Colonel Robert S., 114, 142, 167, 201, 204
Scott, General Winfield, 194
Sharpless, Corporal Edward Clay, 201
Shaw, Sergeant Thomas, 201
Shearin, Sergeant Dan Rogers, 21-23, 30, 52, 61-62, 156
Sheerin, Blacksmith John, 201
Shetter, Cynthia J., 105
Silva, Julian, 116
Simmons-Hogan, Rosie, 146
Simpson, Mayor Howard, 81, 206
Snodgrass, Edward M., 183
Snodgrass, Specialist 4 George E., 63-64, 183-185, 189
Stanley, Private Eben, 137, 201

Stanton, Secretary of War Edwin, 195
Stoeckley, Private First Class Rick, 154-155, 157
Sue, Ray T., 81, 188
Sue, Sergeant Ray E., 63-64, 81, 97, 185, 188-191
Sullivan, Charles, 10, 188
Sullivan, Kenneth "Tay", 10

T

Tabet, State Representative Boni, 113
Tafoya, Mark Alvin, 209
Tanner, Gene P., 24
Taylor, John, 193
Tesillo, Specialist Fourth Class Armando, 61-62, 156
Tripp, State Representative Don, 113

V

Valdez, Marlinda, 133
Valdez, Private First José F., 114, 137, 144, 201, 204
Van Dyke, Master Sergeant (Retired) Gerald, 112
Versace, Humbert Roque, 88
Villareal, Enrique 150

W

Walker, Brigadier General Kenneth, 137, 201
Walker, Doctor Mary Edwards, 198-199
Walker, Representative E.S. "Johnny", 85
Walley, Private Augustus, 201
Washington, General George, 193-194
Wells, Kenneth, 136
Westmoreland, Katherine "Kitsy" (Mrs. William C.), 17

Wilder, Lieutenant Wilbur, 201
Willard, Sergeant First Class Frankie, 22
Williams, Hershel "Woodie", 167-168
Williams, Sergeant Moses, 201
Williams, Sergeant James C., 128
Willis, Hazel Shearin, 21
Wills, Private Henry, 201
Wilson, Jim, 184
Wilson, John Stanton, 209
Woods, Sergeant Brent, 201
Worley, Lance Corporal Kenneth L., 114, 137, 201
Wortman, Henry, 116

Made in the USA
Columbia, SC
05 October 2024